Applied Machine Learning for Healthcare and Life Sciences Using AWS

Transformational AI implementations for biotech, clinical, and healthcare organizations

Ujjwal Ratan

BIRMINGHAM—MUMBAI

Applied Machine Learning for Healthcare and Life Sciences Using AWS

Group Product Manager: Gebin George

Publishing Product Manager: Dinesh Chaudhary

Senior Editor: David Sugarman

Technical Editor: Sweety Pagaria

Copy Editor: Safis Editing

Project Coordinator: Farheen Fathima

Proofreader: Safis Editing

Indexer: Pratik Shirodkar

Production Designer: Shyam Sundar Korumilli

First published: October 2022

Production reference: 1311022

Published by Packt Publishing Ltd.
Livery Place
35 Livery Street
Birmingham
B3 2PB, UK.

ISBN 978-1-80461-021-3

www.packt.com

Contributors

About the author

Ujjwal Ratan is a Principal AI/Machine Learning Solutions Architect at AWS and leads the machine learning solutions architecture group for the AWS healthcare and life sciences industry vertical. Over the years, Ujjwal has been a thought leader in the healthcare and life sciences industry, helping multiple global Fortune 500 organizations achieve their innovation goals by adopting machine learning. His contributions to the AI-driven analysis of medical images, unstructured clinical text, and genomics have helped AWS build products and services that provide highly personalized and precisely targeted diagnostics and therapeutics. Ujjwal's work in these areas has been featured in multiple peer-reviewed publications, technical and scientific blogs, and global conferences.

About the reviewers

Dennis Alvin David is a health informatics and health technology professional and an advocate for the use of data to improve healthcare in general. Coming from a non-traditional background, he worked his way up in the field through a graduate degree in health informatics and attended training in data science, data analytics, data visualization, AI and ML, and data engineering. He strongly believes in continuous learning for personal and professional growth.

Indraneel Chakraborty is a developer working in the biomedical software industry. Passionate about data-driven technical solutions and DevOps, he has previously worked in academic research involving the curation and analysis of clinical trials registry data for insights into policy research. He is also a volunteer maintainer for open source projects, teaching foundational coding and data science skills to professionals and researchers worldwide. He enjoys learning new tech stacks, cloud engineering, and coding.

Table of Contents

Part 2: Machine Learning Applications in the Healthcare Industry

3

4

5

6

Implementing Machine Learning for Medical Devices and Radiology Images   81

Part 3: Machine Learning Applications in the Life Sciences Industry

7

Applying Machine Learning to Genomics   95

8

Applying Machine Learning to Molecular Data   111

9

Applying Machine Learning to Clinical Trials and
Pharmacovigilance   123

10

Utilizing Machine Learning in the Pharmaceutical Supply Chain   143

Part 4: Challenges and the Future of AI in Healthcare and Life Sciences

11

Understanding Common Industry Challenges and Solutions   159

12

Understanding Current Industry Trends and Future Applications   179

Preface

We have seen multiple ways in which AI is touching our lives in a meaningful way. It's almost like having a superpower that you never knew you needed, but now that you have it, you can see how it is changing our lives for the better. Automation driven by AI is helping organizations achieve levels of operational efficiency they never thought were possible, which in turn is boosting economies. The accessibility of cutting-edge infrastructure and models has improved tremendously with the power of cloud computing, which has democratized AI by putting it in the hands of everyone. It is no wonder that the last decade has seen the use of AI in the healthcare and life sciences industry increase massively. As the industry is undergoing a transformation driven by technology and digitization, it produces large volumes of data in multiple modalities. To utilize the full potential of this data, organizations are applying machine learning to process, analyze, and interpret critical information from these datasets to improve and save the lives of patients. It is helping improve provider efficiencies and improve care quality, and is bringing the costs of drugs and therapies down.

This book will help you understand how this is happening. It will introduce you to the different verticals of the healthcare and life sciences industry such as providers, payors, pharmaceuticals, genomics, and medical imaging. It begins by introducing you to the concept of machine learning and then progresses to show how you can apply machine learning to workloads in each of these industry verticals. The book gradually builds your **Amazon Web Services (AWS)** machine learning knowledge. You will be introduced to low-code AI services from AWS and each chapter progresses to more advanced topics. The exercises at the end of the chapters are designed for you to practice what you learned and apply the learning to an actual problem in the industry vertical. I hope you enjoy this ride and find what you learn from this book valuable for a long time to come.

Who this book is for

This book will help you build an understanding of the healthcare and life sciences industry, machine learning and deep learning, and AWS machine learning services. Business and technology decision-makers will see how machine learning is transforming the industry and the role AWS is playing in that. Developers, data scientists, and machine learning engineers will learn about AWS machine learning services and how they can be applied to solve problems in different verticals of the healthcare and life sciences industry. The practical exercises will solidify your knowledge of the concepts learned in each chapter.

What this book covers

Chapter 1, Introducing Machine Learning and the AWS Machine Learning Stack, covers the basic concepts of machine learning and how it differs from a traditional software application.

Chapter 2, Exploring Key AWS Machine Learning Services for Healthcare and Life Sciences, dives into some key machine learning services from AWS that are critical for healthcare and life sciences industries. This chapter will give you an introduction to these services, their key APIs, and some usage examples.

Chapter 3, Machine Learning for Patient Risk Stratification, explains the concept of risk stratification of patients. It shows how common machine learning algorithms for classification and regression tasks can be applied to identify at-risk patients.

Chapter 4, Using Machine Learning to Improve Operational Efficiency for Healthcare Providers, covers operational efficiency in healthcare and why it is important. You will also learn about two common applications of machine learning to improve operational efficiency for healthcare providers.

Chapter 5, Implementing Machine Learning for Healthcare Payors, introduces you to the healthcare payor industry. You will get an understanding of how health insurance organizations process claims.

Chapter 6, Implementing Machine Learning for Medical Devices and Radiology Images, introduces you to the medical device industry. It goes into the details of various regulatory requirements for medical devices to be approved for use based on the type of medical device.

Chapter 7, Applying Machine Learning to Genomics, explores the world of genomes and the evolution of genomic sequencing. We will see how genomic data interpretation and analysis is changing the world of medicine.

Chapter 8, Applying Machine Learning to Molecular Data, introduces molecular data and its interpretation. We will learn about the process of the discovery of new drugs or therapies.

Chapter 9, Applying Machine Learning to Clinical Trials and Pharmacovigilance, covers how we ensure the safety and efficacy of new drugs and therapies before they are available for patients.

Chapter 10, Utilizing Machine Learning in the Pharmaceutical Supply Chain, dives into the world of the pharmaceutical supply chain workflow and introduces you to some challenges in getting new drugs and therapies to patients around the world in a timely manner.

Chapter 11, Understanding Common Industry Challenges and Solutions, summarizes some key challenges, including the regulatory and technical aspects, that deter organizations from adopting machine learning in healthcare and life sciences applications.

Chapter 12, Understanding Current Industry Trends and Future Applications, is all about the future of AI in healthcare and life sciences. We will review some trends in the world of AI/ML and its applications in the healthcare and life sciences industry, understand what's influencing these trends, and see what may lie in store for us in the future.

To get the most out of this book

The exercises in this book require an AWS account and the necessary steps to configure the AWS Python SDK and the AWS **Command Line Interface (CLI)**. You will run the examples from the AWS CLI or a Jupyter notebook from a Sagemaker notebook instance or the Sagemaker Studio environment.

Software/hardware covered in the book	Operating system requirements
Python 3.X	Windows, macOS, or Linux
AWS SDK for Python (Boto 3)	
AWS Command Line Interface (CLI)	

If you are using the digital version of this book, we advise you to type the code yourself or access the code from the book's GitHub repository (a link is available in the next section). Doing so will help you avoid any potential errors related to the copying and pasting of code.

> Note
>
> The exercises in this book use publicly available datasets. Before using these services on any other dataset, please review the AWS HIPPA eligibility guidelines for the respective service. You can learn more about AWS HIPPA guidelines at https://aws.amazon.com/compliance/hipaa-compliance/.
>
> There may be some costs associated with running the example exercises at the end of the chapters. Please follow all best practices around cost optimizations to ensure you are keeping the costs to a minimum. You can learn more about AWS cost optimization at https://aws.amazon.com/architecture/cost-optimization/?cards-all.sort-by=item.additionalFields.sortDate&cards-all.sort-order=desc&awsf.content-type=*all&awsf.methodology=*all.

Download the example code files

You can download the example code files for this book from GitHub at https://github.com/PacktPublishing/Applied-Machine-Learning-for-Healthcare-and-Life-Sciences-using-AWS. If there's an update to the code, it will be updated in the GitHub repository.

We also have other code bundles from our rich catalog of books and videos available at https://github.com/PacktPublishing/. Check them out!

Download the color images

We also provide a PDF file that has color images of the screenshots and diagrams used in this book. You can download it here: https://packt.link/nGhXe.

Conventions used

There are a number of text conventions used throughout this book.

`Code in text`: Indicates code words in text, database table names, folder names, filenames, file extensions, pathnames, dummy URLs, user input, and Twitter handles. Here is an example: "Open the terminal or CLI on your computer and navigate to the directory where you have the `transcribe_text.py` file."

A block of code is set as follows:

```
from sagemaker.workflow.pipeline import Pipeline
pipeline_name = f"mypipeline"
```

When we wish to draw your attention to a particular part of a code block, the relevant lines or items are set in bold:

```
from sagemaker.workflow.parameters import (
    ParameterInteger,
    ParameterString,
    ParameterFloat,
    ParameterBoolean
)
```

Any command-line input or output is written as follows:

```
$cd SageMaker
$git clone https://github.com/PacktPublishing/Applied-Machine-Learning-for-Healthcare-and-Life-Sciences-using-AWS.git
```

Bold: Indicates a new term, an important word, or words that you see onscreen. For instance, words in menus or dialog boxes appear in **bold**. Here is an example: "Select **System info** from the **Administration** panel."

> **Tips or important notes**
> Appear like this.

Get in touch

Feedback from our readers is always welcome.

General feedback: If you have questions about any aspect of this book, email us at customercare@packtpub.com and mention the book title in the subject of your message.

Errata: Although we have taken every care to ensure the accuracy of our content, mistakes do happen. If you have found a mistake in this book, we would be grateful if you would report this to us. Please visit www.packtpub.com/support/errata and fill in the form.

Piracy: If you come across any illegal copies of our works in any form on the internet, we would be grateful if you would provide us with the location address or website name. Please contact us at copyright@packt.com with a link to the material.

If you are interested in becoming an author: If there is a topic that you have expertise in and you are interested in either writing or contributing to a book, please visit authors.packtpub.com.

Share Your Thoughts

Once you've read *Applied Machine Learning for Healthcare and Life Sciences using AWS*, we'd love to hear your thoughts! Scan the QR code below to go straight to the Amazon review page for this book and share your feedback.

https://packt.link/r/1-804-61021-6

Your review is important to us and the tech community and will help us make sure we're delivering excellent quality content.

Download a free PDF copy of this book

Thanks for purchasing this book!

Do you like to read on the go but are unable to carry your print books everywhere? Is your eBook purchase not compatible with the device of your choice?

Don't worry, now with every Packt book you get a DRM-free PDF version of that book at no cost.

Read anywhere, any place, on any device. Search, copy, and paste code from your favorite technical books directly into your application.

The perks don't stop there, you can get exclusive access to discounts, newsletters, and great free content in your inbox daily

Follow these simple steps to get the benefits:

1. Scan the QR code or visit the link below

https://packt.link/free-ebook/9781804610213

2. Submit your proof of purchase
3. That's it! We'll send your free PDF and other benefits to your email directly

Part 1: Introduction to Machine Learning on AWS

In the first part of the book, we will get an overview of AWS machine learning services and see which tools will be a useful part of our journey into healthcare and life sciences.

- *Chapter 1, Introducing Machine Learning and the AWS Machine Learning Stack*
- *Chapter 2, Exploring Key AWS Machine Learning Services for Healthcare and Life Sciences*

1

Introducing Machine Learning and the AWS Machine Learning Stack

Applying **Machine Learning** (**ML**) technology to solve tangible business problems has become increasingly popular among business and technology leaders. There are a lot of cutting-edge use cases that have utilized ML in a meaningful way and have shown considerable success. For example, **computer vision** models can allow you to search for what's in an image by automatically inferring its content, and **Natural Language Processing** (**NLP**) models can understand the intent of a conversation and respond automatically while closely mimicking human interactions. As a matter of fact, you may not even know whether the "entity" on the other side of a phone call is an AI bot or a real person!

While AI technology has a lot of potential for success, there is still a limited understanding of this technology. It is usually concentrated in the hands of a few researchers and advanced partitioners who have spent decades in the field. To solve this knowledge parity, a large section of software and information technology firms such as **Amazon Web Services** (**AWS**) are committed to creating tools and services that do not require a deep understanding of the underlying ML technology and are still able to achieve positive results. While these tools democratize AI, the conceptual knowledge of AI and ML is critical for its successful application and should not be ignored.

In this chapter, we will get an understanding of ML and how it differs from traditional software. We will get an overview of a typical ML life cycle and also learn about the steps a data scientist needs to perform to deploy an ML model in production. These concepts are fairly generic and can be applied to any domain or organization where ML is utilized.

By the end of this chapter, you will get a good understanding of how AWS helps democratize ML with purpose-built services that are applicable to developers of all skill levels. We will go through the AWS ML stack and go over the different categories of services that will help you understand how the AWS AI/ML services are organized overall. We'll cover these topics in the following sections:

- What is ML?

- Exploring the ML life cycle
- Introducing ML on AWS

What is ML?

As the name suggests, ML generally refers to the area of computer science that involves making machines learn and make decisions on their own rather than acting on a set of explicit instructions. In this case, think about the machine as the processor of a computer and the instructions as a program written in a particular programming language. The compiler or the interpreter parses the program and derives a set of instructions that the processor can then execute. In this case, the programmer is responsible for making sure the logic they have in their program is correct as the processor will just perform the task as instructed.

For example, let's assume you want to create a marketing campaign for a new product and want to target the right population to send the email to. To identify this population, you can create a rule in SQL to filter out the right population using a query. We can create rules around age, purchase history, gender, and so on and so forth, and will just process the inputs based on these rules. This is depicted in the following diagram.

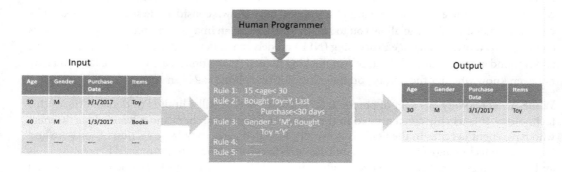

Figure 1.1 – A diagram showing the input, logic, and output of a traditional software program

In the case of ML, we allow the processor to "learn" from past data points about what is correct and incorrect. This process is called **training**. It then tries to apply that learning to unseen data points to make a decision. This process is known as **prediction** because it usually involves determining events that haven't yet happened. We can represent the previous problem as an ML problem in the following way.

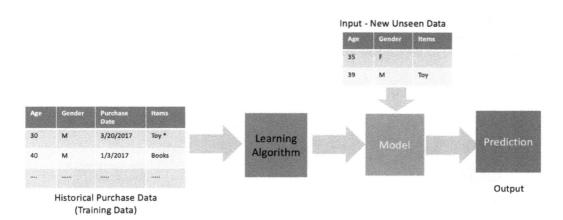

Figure 1.2 – A diagram showing how historical data is used to generate predictions using an ML model

As shown in the preceding diagram, we feed the learning algorithm with some known data points in the form of training data. The algorithm then comes up with a model that is now able to predict outputs from unseen data.

While ML models can be highly accurate, it is worth noting that the output of a model provides a *probabilistic* estimation of an answer instead of *deterministic*, like in the case of a software program. This means that ML models help us predict the probability of something happening, rather than telling us what will happen for sure. For this reason, it is important to continuously evaluate an ML model's output and determine whether there is a need for it to be trained again.

Also, downstream consumers of an ML model (client applications) need to keep the probabilistic nature of the output in mind before making decisions based on it. For example, software to compute the sales numbers at the end of each quarter will provide a deterministic figure based on which you can calculate your profit for the quarter. However, an ML model will predict the sales number at the end of a future quarter, based on which you can predict what your profit would look like. The former can be entered in the books or ledger but the latter can be used to get an idea of the future result and take corrective actions if needed.

Now that we have a basic understanding of how to define ML as a concept, let's look at two broad types of ML.

Supervised versus unsupervised learning

At a high level, ML models can be divided into two categories:

- **Supervised learning model**: A supervised ML model is created when the training data has a target variable in place. In other words, the training data contains unique combinations of input features and target variables. This is known as a labeled dataset. The supervised learning model learns the relationship between the target and the input features during the training process. Hence, it is important to have high-quality labeled datasets when training supervised learning models. Examples of supervised learning models include classification and regression models. *Figure 1.3* depicts how this would work for a model that recognizes the breed of dogs.

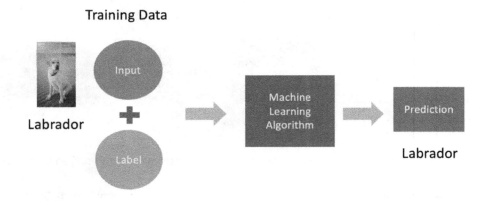

Figure 1.3 – A diagram showing a typical supervised learning prediction workflow

- **Unsupervised learning model:** An unsupervised learning model does not depend on the availability of labeled datasets. In other words, unlike its supervised cousin, the unsupervised learning model does not learn the association between the target variable and the input features. Instead, it learns the patterns in the overall data to determine how similar or different each data point is from the others. This is usually done by representing all the data points in parameter space in two or three dimensions and calculating the distance between them. The closer they are to each other, the more similar they are to each other. A common example of an unsupervised learning model is **k-means clustering**, which divides the input data points into *k* number of groups or clusters.

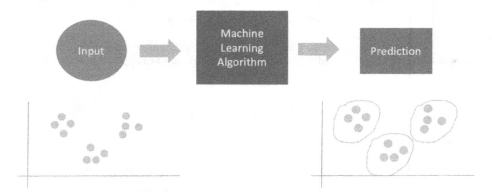

Figure 1.4 – A diagram showing a typical unsupervised learning prediction workflow

Now that we have an understanding of the two broad types of ML models, let us review a few key terms and concepts that are commonly used in ML.

ML terminology

There are some key concepts and terminologies specific to ML and it's important to get a good understanding of these concepts before you go deeper into the subject. These terms will be used repeatedly throughout this book:

- **Algorithm**: Algorithms are at the core of an ML workflow. The algorithm defines how the training data is utilized to learn from its representations, and is then used to make predictions of a target variable. An example of an algorithm is **linear regression**. This algorithm is used to find the best fit line that minimizes the error between the actual and the predicted values of the target. This best fit line can be represented by a linear equation such as y=ax+b. This type of algorithm can be used for problems that can be represented by linear relationships. For example, predicting the height of a person based on their age or predicting the cost of a house based on its square footage.

  However, not all problems can be solved using a linear equation because the relationship between the target and the input data points might be **non-linear**, represented by a curve instead of a straight line. In the case of non-linear regression, the curve is represented by a nonlinear equation y=f(x,c)+b where f(x,c) can be any non-linear function. This type of algorithm can be used for problems that can be represented by non-linear relationships. For example, the prediction of the progression of a disease in a population can be driven by multiple non-linear relationships. An example of a non-linear algorithm is a **decision tree**. This algorithm aims to learn how to split the data into smaller subsets till the subset is as close in representation to the target variable as possible.

  The choice of algorithm to solve a particular problem ultimately depends on multiple factors. It is often recommended to try multiple algorithms and find the one that works best for a particular problem. However, having an intuition of how the algorithm works allows you to narrow it down to a few.

- **Training**: Training is the process by which an algorithm learns. In other words, it helps converge on the best fit line or curve based on the input dataset and the target variable. As a result, it is also sometimes referred to as fitting. During the training process, the input data and the target are fed into the algorithm iteratively, in batches. The process tries to determine the coefficients of the final equation that represents the line or the curve with the minimum error when compared to the target variable. During the training process, the input dataset is divided into three different groups: train, validation, and test. The train dataset is the majority of the input data and is used to fit or train the model. The validation dataset is used to evaluate the model performance and, if necessary, tune the model input parameters (also known as **hyperparameters**) in an iterative way. Finally, the test dataset is the dataset on which the final evaluation of the model is done, which determines whether it can be deployed in production.

The process of training and tuning the model is highly iterative. It requires multiple runs and trial and error to determine the best combination of parameters to use in the final model.

The evaluation of the ML model is done by a metric, also known as the **evaluation metric**, which determines how good the model is. Depending on the choice of the evaluation metric, the training process aims to minimize or maximize the value of the metric. For instance, if the evaluation metric is **Mean Squared Error**, the goal of the training job is to minimize it. However, if the evaluation metric is accuracy, the goal would be to maximize it. Training is a compute-intensive process and can consume considerable resources.

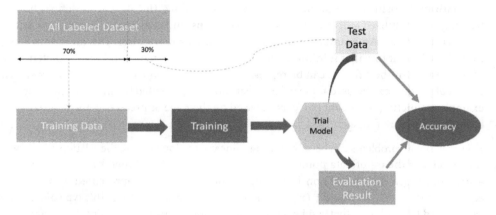

Figure 1.5 – A diagram showing the steps of a model training workflow

- **Model**: An ML model is an artifact that results from the training process. In essence, when you train an algorithm with data, it results in a model. A model accepts input parameters and provides the predicted value of the target variable. The input parameters should be exactly the same in structure and format as the training data input parameters. The model can be serialized into a format that can be stored as a file and then deployed into a workflow to generate predictions. The serialized model file stores the weights or coefficients that, when applied to the equation, result in the value of the predicted target. To generate predictions, the model needs to be de-serialized or reconstructed from the saved model file. The idea of saving the model to the disk by serializing it allows for model **portability**, a term typically used by data scientists to denote interchangeability between frameworks when it comes to data science. Common ML frameworks such as **PyTorch**, **scikit-learn**, and **TensorFlow** all support serializing model files into standard formats, which allows you to standardize your model registries and also use them interchangeably if needed.

- **Inference**: Inference is the process of generating predictions from the trained model. Hence, this step is also known as **predicting**. The model that has been trained on past data is now exposed to unseen data to generate the value of the target variable. As described earlier, the model resulting from the training process is already evaluated using a metric on a test dataset, which is a subset of the input data. However, this does not guarantee that the model will perform well on unseen data when deployed. As a result, prediction results are continuously monitored and compared against the ground truth (actual) values.

There are various ways in which the model results are monitored and evaluated in production. One common method utilizes humans to evaluate certain prediction results that are suspicious. This method of validation is also known as **human-in-the-loop**. In this method, the model results with lower confidence (usually denoted by a **confidence score**) are routed to a human to determine if the output is correct or not. The human can then override the model result if needed before sending it to the downstream system. This method, while extremely useful, has a drawback. Some ML models do not have the ground truth data available until the event actually happens in the future. For instance, if the model predicts a patient is going to require a kidney transplant, a human may not be able to validate that output of the model until the transplant actually happens (or not). In such cases, the human-in-the-loop method of validation does not work. To account for such limitations, the method of monitoring drift in real-world data compared to the training data is utilized to determine the effectiveness of the model predictions. If the real-world data is very different from the training data, the chances of predictions being correct are minimal and hence, it may require retraining the model.

Inferences from an ML model can be executed as an **asynchronous batch process** or as a **synchronous real-time process**. An asynchronous process is great for workloads that run in batches. For example, calculating the risk score of loan defaults across all monthly loan applications at the end of the month. This risk score is generated or updated once a month for a large batch of individuals who applied for a loan. As a result, the model does not need to serve requests 24/7 and is only used at scheduled times. Synchronous or real-time inference is when the model serves out inference as a response to each request 24/7 in real time. In this case, the model needs to be hosted on a highly available infrastructure that remains up and running at all times and also adheres to the latency requirements of the downstream application. For example, a weather forecast application that continuously updates the forecast conditions based on the predictions from a model needs to generate predictions 24/7 in real time.

Now that we have a good understanding of what ML is and the key terminologies associated with it, let's look at the process of building the model in more detail.

Exploring the ML life cycle

The **ML life cycle** refers to the various stages in the conceptualization, design, development, and deployment of an ML model. These stages in the ML model development process consist of a few key steps that help data scientists come up with the best possible outcome for the problem at hand. These steps are usually repeatable and iterative and are combined into a pipeline commonly known as the ML pipeline. An ideal ML pipeline is automated and repeatable so it can be deployed and maintained as a production pipeline. Here are the common stages of an ML life cycle.

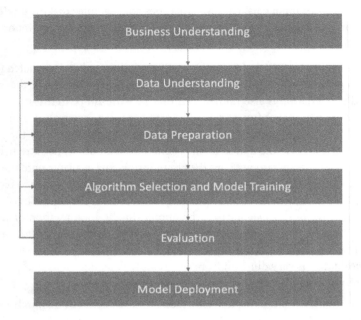

Figure 1.6 – A diagram showing the steps of an ML life cycle

Figure 1.6 shows the various steps of the ML life cycle. It starts with having a business understanding of the problem and ends with a deployed model. The iterative steps such as data preparation and model training are denoted by loops to depict that the data scientists would perform those steps repeatedly until they are satisfied with the results. Let us now look at the steps in more detail.

Problem definition

A common mistake is to think ML can solve any problem! Problem definition is key to determining whether ML can be utilized to solve it. In this step, data scientists work with business stakeholders to find out whether the problem satisfies the key tenets of a good ML problem:

- **Predictive element**: During the ML problem definition, data scientists try to determine whether the problem has a predictive element. It may well be the case that the output being requested can be modeled as a rule that is calculated using existing data instead of creating a model to predict it.

    For example, let us take into consideration the problem of health insurance claim fraud identification. There are some tell-tale signs of a claim being fraudulent that are derivable from the existing claims database using data transformations and analytical metrics. For example, verifying whether it's a duplicate claim, whether the claim amount is unusually high, whether the reason for the claim matches the patient demographic or history, and so on.

These attributes can help determine the high-risk claim transactions, which can then be flagged. For this particular problem, there is no need for an ML model to flag such claim transactions as the rules applied to existing claim transaction data are enough to achieve what is needed. On the other hand, if the solution requires a deeper analysis of multiple sources of data and looks at patterns across a large volume of such transactions, it may not be a good candidate for rules or analytical metrics. Applying conventional analytics to large volumes of heterogeneous datasets can result in extremely complicated analytical queries that are hard to debug and maintain. Moreover, the processing of rules on these large volumes of data can be compute-intensive and may become a bottleneck for the timely identification of fraudulent claims. In such cases, applying ML can be beneficial. A model can look at features from different sources of data and learn how they are associated with the target variable (fraud versus no fraud). It can then be used to generate a risk score for each new claim.

It is important to talk to key business stakeholders to understand the different factors that go into determining whether a claim is fraudulent or not. In the process, data scientists document a list of input features that can be used in the ML model. These factors help in the overall determination of the predictive element of the problem statement.

- **Availability of dataset**: Once the problem is determined to be a good candidate for ML, the next important thing to check is the availability of a high-quality labeled dataset. We cannot train models without the available data. The dataset should also be clean, with no missing values, and be evenly distributed across all features and values. It should have mapping to the target variable and the target itself should be evenly distributed across the dataset. This is obviously the ideal condition and real-world scenarios may be far from ideal. However, the closer we can get to this ideal state of data, the easier it is to produce a highly accurate model from it. In some cases, data scientists may recommend to the business they collect more data containing more examples of a certain type or even more features before starting to experiment with ML methods. In other cases, labeling and annotation of the raw data by **subject matter experts (SMEs)** may be needed. This is a time-consuming step and may require multiple rounds of discussions with the SMEs, business stakeholders, and data scientists before arriving at an appropriate dataset to begin the ML modeling process. It is worth the time, as utilizing the right dataset ensures the success of the ML project.

- **Appetite for experimentation**: In a few scenarios, it is important to highlight the fact that data science is a process of experimentation and the chances of it being successful are not always high. In a software development exercise, the work involved in each phase of requirements gathering, development, testing, and deployment can be largely predictable and can be used to accurately estimate the time it will take to complete the project. In an ML project, that may be difficult to determine from the outset. Steps such as data gathering and training the tuning hyperparameters are highly iterative and it could take a long time to come up with the best model. In some cases where the problem and dataset are well known, it may be easier to estimate the time as the results have been proven. However, the time taken to solve novel problems using ML methods could be difficult to determine. It is therefore recommended that the key stakeholders are aware of this and make sure the business has an appetite for experimentation.

Data processing and feature engineering

Before data can be fed into an algorithm for training a model, it needs to be transformed, cleaned, and formatted in a way that can be understood by ML algorithms. For example, raw data may have missing values and may not be standardized across all columns. It may also need transformations to create new derived columns or drop a few columns that may not be needed for ML. Once these data processing steps are complete, the data needs to be made suitable for ML algorithms for training. As you know by now, algorithms are representative of a mathematical equation that accepts the input values of the training datasets and tries to learn its association with the target. Therefore, it cannot accept non-numeric values. In a typical training dataset, you may have numeric, categorical, or text values that have to be appropriately engineered to make them appropriate for training. Some of the common techniques of feature engineering are as follows:

- **Scaling**: This is a technique by which a feature that may vary a lot across the dataset can be represented at a common scale. This allows the final model to be less sensitive to the variations in the feature.

- **Standardizing**: This technique allows the feature distribution to have a mean value of zero and a standard deviation of one.

- **Binning**: This approach allows for granular numerical values to be grouped into a set, resulting in categorical variables. For example, people above 60 years of age are old, between 18 and 60 are adults, and below 18 are young.

- **Label encoding**: This technique is used to convert categorical features into numeric features by associating a numerical value to each unique value of the categorical variable. For example, if a feature named `color` consists of three unique values – `Blue`, `Black`, and `Red` – label encoders can associate a unique number with each of those colors, such as `Blue=1`, `Black=2`, and `Red=3`.

- **One-hot encoding**: This is another technique for encoding categorical variables. Instead of assigning a unique number to each value of a categorical feature, this technique converts each feature into a column in the dataset and assigns it a 1 or 0. Here is an example:

Price	Model
1000	iPhone
800	Samsung
900	Sony
700	Motorola

Table 1.1 – A table showing data about cell phone models and their price

Applying one-hot encoding to the preceding table will result in the following structure.

Price	iPhone	Samsung	Sony	Motorola
1000	1	0	0	0
800	0	1	0	0
900	0	0	1	0
700	0	0	0	1

Table 1.2 – A table showing the results of one-hot encoding applied to the table in Figure 1.7

The resulting table is sparse in nature and consists of numeric features that can be fed into an ML algorithm for training.

The data processing and feature engineering steps you ultimately apply depend on your source data. We will look at some of these techniques applied to datasets in subsequent chapters where we will see examples of building, training, and deploying ML models with different datasets.

Model training and deployment

Once the features have been engineered and are ready, it is time to enter into the training and deployment phase. As mentioned earlier, it's a highly repetitive phase of the ML life cycle where the training data is fed into the algorithm to come up with the best fit model. This process involves analyzing the output of the training metrics and tweaking the input features and/or the hyperparameters to achieve a better model. Tuning the hyperparameters of a model is driven by intuition and experience. Experienced data scientists select the initial parameters based on their knowledge of solving similar problems using the algorithm of choice and can come up with the best fit model faster. However, the trial-and-error process can be time-consuming for a new data scientist who is starting off with a random search of the parameters. This process of identifying the best hyperparameters of a model is known as hyperparameter tuning.

The trained model is then deployed typically as a REST API that can be invoked for generating predictions. It's important to note that training and deployment is a continuous process in an ML life cycle. As discussed earlier, models that perform well in the training phase may degrade in performance in production over a period of time and may require retraining. It is also important to keep training the model at regular intervals with newly available real-world data to make sure it is able to predict accurately in all variations of production data. For this reason, ML engineers prefer to create a repeatable ML pipeline that continuously trains, tunes, and deploys newer versions of models as needed. This process is known as **ML Operations**, or simply **MLOps**, and the pipeline that performs these tasks is known as an **MLOps pipeline**.

Introducing ML on AWS

AWS puts ML in the hands of every developer, irrespective of their skill level and expertise, so that businesses can adopt the technology quickly and effectively. AWS focuses on removing the undifferentiated heavy lifting in the process of building ML models such as the management of the underlying infrastructure, the scaling of the training and inference jobs, and ensuring high availability of the models. It provides developers with a variety of compute instances and containerized environments to choose from that are purpose-built for the accelerated and distributed computing needed for high-scale ML jobs. AWS has a broad and deep set of ML capabilities for builders that can be connected together, like Lego pieces, to create intelligent applications.

AWS ML services cover the full life cycle of an ML pipeline from data annotation/labeling, data cleansing, feature engineering, model training, deployment, and monitoring. It has purpose-built services for problems in computer vision, natural language processing, forecasting, recommendation engines, and fraud detection, to name a few. It also has options for automatic model creation and no-/low-code options for creating ML models. The AWS ML services are organized into three layers also known as the **AWS machine learning stack**.

Introducing the AWS ML stack

The following diagram represents the version of the AWS AI/ML services stack as of April 2022.

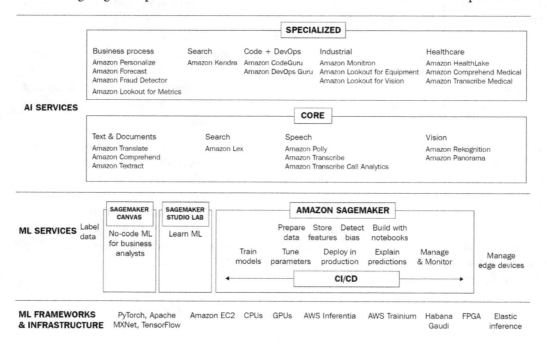

Figure 1.7 – A diagram depicting the AWS ML stack as of April 2022

The stack can be used by expert practitioners who want to develop a project within the framework of their choice; data scientists who want to use the end-to-end capabilities of SageMaker; business analysts who can build their own model using Canvas; or application developers with no previous ML skills who can add intelligence to their applications with the help of API calls. The following are the three layers of the AWS AI/ML stack:

- **AI services layer**: The AI services layer of the AWS ML stack is the topmost layer of the stack. It consists of services that require minimal knowledge of ML. Sometimes, it comes with a pre-trained model that can be just invoked using APIs from the AWS SDK, the AWS CLI, or the console. In other cases, the services allow you to customize the model by providing your own labeled training dataset so the responses are more appropriate for the problem at hand. In any case, the AI services layer of the AWS AI/ML stack is focused on ease of use. The services are designed for specialized applications in industrial settings, search, business processes, and healthcare. It also comes with a core set of capabilities in the areas of speech, chatbots, vision, and text and documents.

- **ML services layer**: The ML services layer is the middle layer of the AWS AI/ML stack. It provides tools for data scientists to perform all the steps of the ML life cycle, such as data cleansing, feature engineering, model training, deployment, and monitoring. It is driven by the core ML platform of AWS known as **Amazon SageMaker**. SageMaker provides the ability to build a modular containerized environment that interfaces with the AWS compute and storage services seamlessly. It provides its own SDK that has APIs to interact with the service. It removes the complexity from each step of the ML workflow by providing simple-to-use modular capabilities with a choice of deployment architectures and patterns to suit virtually any ML application. It also contains MLOps capabilities to create a reproducible ML pipeline that is easy to maintain and scale. The ML services layer is suited for data scientists who build and train their own models and maintain large-scale models in production environments.

- **ML fameworks and the infrastructure layer**: The ML frameworks and infrastructure layer is the bottom layer of the AWS AI/ML stack. The services in this layer are for expert practitioners who can develop using the framework of their choice. It provides a choice for developers and scientists to run their workloads as a managed experience in Amazon SageMaker or run their workloads in a self-managed environment on AWS Deep Learning, **Amazon machine images** (**AMIs**), and AWS Deep Learning Containers. The AWS Deep Learning AMI and containers are fully configured with the latest versions of the most popular deep learning frameworks and tools – including PyTorch, MXNet, and TensorFlow. As part of this layer, AWS provides a broad and deep portfolio of compute, networking, and storage infrastructure services with a choice of processors and accelerators to meet your unique performance and budget needs for ML.

Now that we have a good understanding of ML and the AWS ML stack, it is a good time to re-read sections that may not be entirely clear. Also, the chapter introduces concepts of ML, but if you want to dive deeper into any of the concepts touched upon in this chapter, there are several trusted online resources for you to refer to. Let us now summarize the lessons from this chapter and see what's ahead.

Summary

In this chapter, you got an overview of the basic concepts of ML. You went over the definition of ML and how it differs from typical software. You also learned about important terminologies and concepts that are heavily used in the context of ML. The chapter also covered the important steps of the ML life cycle, which can be combined together to create an end-to-end ML pipeline to deploy models in production. Lastly, you got an introduction to the AWS ML stack and how the AWS AI/ML services are organized.

In *Chapter 2, Exploring Key AWS Machine Learning Services for Healthcare and Life Sciences*, we will dive into the details of some of the critical AWS services that allow healthcare and life sciences customers to build, train, and deploy ML models for solving important problems in the industry. We will cover those problems in detail in the subsequent chapters of this book.

2

Exploring Key AWS Machine Learning Services for Healthcare and Life Sciences

AWS is the leader in cloud computing technology. According to the Gartner report published in July 2021 that covers cloud infrastructure and platform services, they recognize AWS as a leader placed highest in the Magic Quadrant, a mechanism used by Gartner to categorize candidates in their evaluation report. They rank it highest in both axes of measurement, the ability to execute and the completeness of vision. AWS also provides one of the most complete and feature-rich lists of services for ML.

The vast number of services on AWS might be confusing for someone just starting to build with AWS. Specifically, healthcare and life sciences organizations that have requirements that cater only to the domain may find it difficult to operate with domain agnostic services that AWS provides. To facilitate healthcare and life sciences organizations running ML workloads on AWS, AWS has created a set of services that are specifically designed for solving business problems in the healthcare and life sciences industry. In other cases, there are features that are available in domain-agnostic services such as Amazon SageMaker that can be easily tailored for specific healthcare and life science workflows. These services provide the necessary controls for security, privacy, and regulatory requirements that are a prerequisite for any healthcare and life sciences organization.

In this chapter, we will get into the details of why ML is needed in healthcare and life sciences and learn about some of the key ML use cases that healthcare and life sciences organizations are building today to create a positive impact in the industry. Then, we will revisit the AWS ML stack and dive deeper into some key AWS AI/ML services designed specifically for the healthcare and life sciences domain. We will also look at some domain-agnostic AI/ML services that can be used for healthcare and life sciences workloads because of the capability they provide to end users.

By the end of this chapter, you will get a deeper understanding of some key AWS services such as Amazon Comprehend Medical, Amazon Transcribe Medical, and Amazon HealthLake. We will cover the following topics:

- Applying ML in healthcare and life sciences

- Introducing AWS AI/ML services for healthcare and life sciences
- Introducing Amazon HealthLake

Applying ML in healthcare and life sciences

In the first chapter, we went over the introductory concepts of ML and how it differs from typical software. We also covered the ML life cycle and the different steps involved in an ML project. Let us now apply this understanding to healthcare and life sciences and look at some examples of how ML is impacting the healthcare and life sciences industry.

The healthcare and life sciences industry can be divided into multiple subsegments that organizations help support. It starts from research that allows for the discovery of new therapeutics and drugs and helps understand the human body by mapping the genetic code. It includes the process of taking drugs into clinical trials and tracking their progress and regulatory reporting through various stages of testing to ascertain whether the drug is safe. It involves manufacturing the drugs at scale for fast global distribution and matching that with targeted sales and commercial campaigns to ensure the maximum success of the drug. It involves making sure care providers have all the necessary tools to provide the best possible care and prescribe the right medications to targeted populations. Finally, it involves the reimbursement of care providers, taking into account the goal of reducing the cost of healthcare while increasing care quality.

Machine learning has a big part to play in each of these areas and has allowed healthcare and life sciences organizations to improve the way they run their businesses. We will cover some specific examples in each of these areas in detail in the subsequent chapters of this book. For now, let us get an overview of some use cases for each segment.

Healthcare providers

Healthcare providers are licensed to diagnose and treat medical conditions. It can be an individual physician or a facility such as a clinic or a hospital. The performance of a healthcare provider is **determined** by their ability to provide high care quality while keeping costs low. These performance metrics determine how healthcare providers are paid for their services and hence play a very important role in the overall success of the provider's business.

One area in that ML has had a huge impact on care quality is by identifying patients in the population that are at a high risk of a negative outcome. This could be a risk of mortality or a risk of a worsening medical condition. Another common risk for healthcare providers is the risk of readmission. It occurs when a patient discharged from the facility is admitted again to the hospital within a set number of days (typically 30 days). Healthcare providers are tracked for their rate of readmissions. ML can identify at-risk patients early so that necessary interventions can be done while the patient is in the hospital, which will help prevent readmission and keep the rate of readmission low.

Another common area of the application of ML for healthcare providers is operational efficiency. Healthcare providers are under a lot of pressure, which causes provider burnout and results in a negative impact on their ability to provide critical care. Providers spend considerable time taking notes and doing back-office work that takes time away from patient care and results in poor efficiency. ML can help automate manual steps that result in better operational efficiency and improved care quality for healthcare providers.

Healthcare payors

Healthcare payors are in charge of paying the providers for the services they offer and processing their claims. Healthcare payors design health plans containing service rates for healthcare services that subscribers in the plan can claim. For example, in the US, Medicare is a national health insurance service for people over 65 years of age. Healthcare payors need to process claims fast and efficiently to ensure they are able to meet their volumetric requirements and maintain their profit margins. At the same time, they need to ensure they process claims correctly to avoid any waste of money and resources.

ML can help with both these tasks. Healthcare payors have been utilizing ML to identify fraudulent claims so they can be flagged and followed up on, saving them a lot of time dealing with unwarranted claims. Moreover, the adjudication of claims involves several manual steps such as processing the claim forms and checking subscriber coverage based on their enrolment in a plan. These are steps that ML models can automate. This helps healthcare payors process a larger volume of claims in the same amount of time. Lastly, ML models can learn from past claims data and identify areas of optimization in health plans that can save money. These include identification of costly procedures that can be avoided by proactive steps such as preventive care and also identification of individuals who may be at a high risk of a disease that could be very costly and deciding appropriate mechanisms to offset that cost such as higher premiums.

Medical devices and imaging

The term **medical device** is used to describe a large category of medical equipment and consumables that aid in activities such as treatment, diagnosis, and cure of certain diseases and chronic conditions. For example, a **Magnetic Resonance Imaging (MRI)** machine is a highly specialized medical device used by radiologists to generate pictures of human anatomy that can be used to help diagnose a variety of cancers. Contact lenses are also medical devices that are commonly worn to correct vision in lieu of prescription glasses. In the US, the **Food and Drug Administration (FDA)** classifies medical devices into three classes. *Class 1* medical devices are low risk and include consumables such as bandages and some handheld surgical instruments. A *class 2* medical device is an intermediate-risk device and includes devices such as pumps and scanners. A *class 3* medical device is a high-risk device. It is critical to sustaining life and is highly regulated. Examples include pacemakers and defibrillators.

The application of ML in medical devices is a developing topic and is challenging due to a variety of regulations that go along with making any changes to the underlying operations of the device. A device needs to be *validated* for use by the FDA and any change in its operating process or design process needs a revalidation. Hence, using technologies such as ML is a difficult proposition. However, the FDA understands the value ML can bring to medical devices and created guidance for medical device organizations to create validated ML pipelines.

One common application of ML to medical devices is in the space of smart connected devices. For example, smart glucometers can take regular readings of blood glucose levels and stream that to the cloud for aggregation. You can then analyze the trends in glucose readings for an individual and forecast the onset of chronic diabetes or other diseases. You can also recommend and alert individuals to adhere to their medications and care plan so they can avoid serious medical conditions in the future. ML models can also run directly on these medical devices, also known as **ML models on the edge**. For example, ML models to segment parts of a tumor in an X-ray can be trained on AWS and then deployed on X-ray machines to run locally and highlight tumors in X-rays for easy identification and diagnosis.

Pharmaceutical organizations

Pharmaceutical organizations have licenses to develop, test, market, and distribute drugs to treat medical conditions. They also conduct research to develop new drugs and therapies. The pharmaceutical industry works with multiple partners who help them with conducting clinical trials, manufacturing drugs, and distributing them globally. For example, they work with **Clinical Research Organizations (CROs)** that conduct extensive clinical trials for the drugs. In fact, the process of taking a drug to market can take 10 years and require billions of dollars in investments. Hence, it's really important to market the drugs appropriately to maximize the return on investment.

ML can help optimize the process of taking a drug to market and help save money in the process. For example, the process of drug discovery can involve searching across billions of molecules using trial and error. ML algorithms can dramatically reduce this search space by identifying probable targets. In the clinical trial process, ML can help with identifying the right trial participants based on the sites where the trials would be conducted and the inclusion and exclusion criteria of the trials defined in the trial protocol. It can help detect and report adverse drug reactions during the trial process. ML models can also help with predicting critical equipment failures before they happen to prevent any bottlenecks in the manufacturing of the drugs. It can help forecast demand and help with logistics and planning around the distribution of the drug in particular regions and target markets. It can also track drug reviews from healthcare providers and consumers (patients) to understand sentiments around the drug and take corrective action to address negative reviews that may jeopardize sales and profits.

Genomics

Genomics is the branch of biology that works in the area of understanding the function of genomes, the complete DNA of an organism. In the case of the human genome, we are still early in the journey of the understanding and full-scale adoption of genomics-driven clinical interventions. However, the successful mapping of the whole human genome in 2003 led to a slew of research that has fueled discoveries of novel genes. DNA is made up of four types of nucleic acids, namely: T (thymine), C (cytosine), G (guanine), and A (adenine), that occur in a particular sequence. It looks like a twisted ladder that forms the iconic double helix that is normally used to portray DNA graphically. A majority of genes (99.9%) in all humans are arranged in the same manner. The 0.1% variation in this order is enough to define the unique characteristics of an individual such as the color of your eyes or skin. These variations, known as genetic variants, sometimes can put an individual at a higher risk of certain diseases and decide their response to certain types of medications. The knowledge of these variants and how they correlate with certain diseases are the basis of the field of **precision medicine**. This field is centered around the design of drugs and therapies that are unique to an individual's genetic code instead of mass-producing drugs that are the same for everyone. It is shown in research that individualized care plans and therapies driven by genomic data for the treatment of serious diseases such as cancer produce better outcomes for patients.

The identification of these unique genes and variants requires sequencing of the human genome and processing the resulting data. This process is known as **Next-Generation Sequencing** (**NGS**). The cost of whole genome sequencing has reduced considerably in the last decade, from thousands of dollars to a few hundred now. This has led to large population-level sequencing initiatives by several countries. For example, the UK biobank project consists of genetic sequences and health information for half a million participants from the UK. These types of large-scale sequencing studies aim to identify genetic variation and answer evolutionary questions at a population level. The resulting data produced is in the petabyte scale and needs specialized technology to store, process, and interpret it for clinical relevance.

One common area where ML has been applied successfully in genomics is the area of correlation studies. When it comes to genetic variants, it is important to understand their effects on certain types of diseases and other observable characteristics of an individual, known as phenotypes. For example, ML can help identify patients with a high risk of breast cancer by analyzing their gene expressions. It can also identify individuals with lactose intolerance or other types of allergies.

Now that we have seen the applications of ML in different healthcare and life sciences segments, let us look at some key AWS offerings that allow you to build these applications.

Introducing AWS AI/ML services for healthcare and life sciences

The AWS AI/ML stack consists of purpose-built services that can help organizations develop ML applications for a variety of use cases. Some of these services are unique to healthcare and life sciences and are designed to solve specific industry problems. Others are not designed specifically for healthcare and life sciences but are a great building block for ML applications in healthcare and life sciences. Let us look at some of these services in more detail.

Introducing Amazon Comprehend Medical

Amazon Comprehend Medical is the first healthcare-specific ML service from AWS. It falls in the category of **Natural Language Processing** (**NLP**), a branch of ML that allows models to understand the nuances of human language. Amazon Comprehend Medical extends this concept to the healthcare domain, by helping understand medically relevant information from unstructured clinical text. The service performs **Named Entity Extraction** (**NEE**) of key medical entities from texts, such as discharge summaries, progress notes, visit summaries, and a variety of other unstructured clinical notes. The service provides pre-trained models that are accessible using the **AWS SDK** via APIs, the **AWS Command Line Interface** (**CLI**), and the **AWS console**. It is a *stateless* service, which means it does not store any end-user information passed to it. It also falls under the category of **AWS HIPAA Eligible Services**. These services allow you to store, process, and transmit **Protected Health Information** (**PHI**) information.

Amazon Comprehend Medical provides easy-to-use APIs. These APIs provide an output in JSON format that is easy to process and integrate with downstream applications or even store in a database. These APIs can be called in a synchronous manner or an asynchronous manner. You can choose between synchronous or asynchronous processing based on your overall application architecture. The functionalities that the service provides include the following:

- Extract clinical entities by performing NER on unstructured clinical notes. There are various categories of entities that the services can recognize. The entities are automatically detected from unstructured clinical text using ML models. The following are the categories of entities that the service recognizes:

  - **Anatomy**: References to parts of a body and their location

  - **Medical condition**: For example, signs, symptoms, and diagnosis

  - **Medication**: Medicine names and dosage information

  - **Test treatment procedure**: Procedures that determine medical conditions

  - **Time expression**: Detects time elements in the notes when associated with an entity.

  - **Protected health information**: Detects PHI elements in the notes

Amazon Comprehend Medical also provides a dedicated PHI detection functionality, which is a subset of the larger entity extraction capability. This alternative is ideal for anyone looking to just detect sensitive information in medical notes and who does not want to get visibility into the full list of entities. The PHI detection feature can be used for the de-identification of medical notes, a process of removing or masking PHI information.

In addition to identifying entities, the service also detects attributes and their relationships to the entities. For example, the dosage is an attribute of medication and is related to it automatically by the service. It also recognizes *traits* of an entity. Traits contain information about the entity that Amazon Comprehend Medical detects from context. For example, **NEGATION** is a trait that can be detected by Comprehend Medical and attached to the medication entity when the patient is not taking the medication.

For each of the entities, the output provides a **confidence score**, which is a measure of how sure the model is of the prediction being made. It also provides the beginning and end offset of the text, which denotes the position where the text occurs.

- Amazon Comprehend Medical can detect the 2021 version of the international classification of diseases, 10th revision, clinical modification (ICD-10-CM) codes provide by the **Centers for Disease Control and Prevention (CDC)**. ICD-10-CM codes are associated with medical conditions and are used extensively for a variety of medical processes such as ordering procedures and making reimbursements. For each medical condition detected by Amazon Comprehend Medical, there are five ICD-10-CM codes provided in decreasing order of confidence.

- Amazon Comprehend Medical can detect the **concept identifiers** (RxCUI) from the RxNorm database from the National Library of Medicine. The source for each RxCUI is the UMLS Metathesaurus 2020AB. The RxNorm codes are used for identifying medications and standardizing their representations across different healthcare systems such as billing and ordering systems. It is also used in filling and ordering prescriptions.

- Amazon Comprehend Medical can detect concepts from the 2021-03 version of the Systematized Nomenclature of Medicine, Clinical Terms (SNOMED CT). SNOMED codes are widely used for standardizing medical information and provide a comprehensive vocabulary of medical information. It covers entities such as medical conditions, tests, treatments, procedures, and anatomy. Just like in the case of ICD-10-CM and the RxNorm ontology, Amazon Comprehend Medical provides the top five SNOMED codes for the detected medical concept in decreasing order of confidence. The codes can be used to standardize medical information and automate the process of manual coding. Let us look at an example of the SNOMED code detection feature of Amazon Comprehend Medical.

Amazon Comprehend Medical can be used as a building block for a variety of healthcare and life sciences applications that depend on processing unstructured information. The APIs of Comprehend Medical described in this section can also be combined with other AWS services to create a workflow for use cases such as population health management, medical billing and claims processing, and medical coding. You can refer to the developer guide of Amazon Comprehend Medical for more details on how to use these APIs for your applications.

A very common workflow is to integrate Amazon Transcribe Medical with Amazon Comprehend Medical for medical transcription and conversation analysis. Let us now look at Amazon Transcribe Medical in more detail.

Introducing Amazon Transcribe Medical

Amazon Transcribe Medical is an **automatic speech recognition (ASR)** service specifically designed for the medical domain. Just like **Amazon Transcribe**, which is a generic version of the service, Amazon Transcribe Medical covers speech-to-text and transcribes a variety of clinical terminology such as medication names, disease names, and medical procedures. The service covers primary care and also specializations such as cardiology, neurology, obstetrics-gynecology, pediatrics, oncology, radiology, and urology.

Amazon Transcribe Medical is a HIPAA-eligible service and provides a pretrained model that can be accessed via different APIs available in multiple programming languages. When combined with Amazon Comprehend Medical, you can create an end-to-end workflow for a variety of applications in healthcare such as analyzing doctor-patient exchanges, detecting medical entities in real time during a telehealth visit, or summarizing patient visit information for recording into the **Electronic Medical Record** (**EMR**) system.

Amazon Transcribe Medical can be used to transcribe audio in conversation format or dictation format:

- **Conversation**: The conversation option allows you to transcribe interactions between different individuals. For example, an interaction between a doctor and a patient. In this mode, the service allows you to identify and label each speaker in the conversation. This is useful when trying to identify what was said by the doctor and what was said by the patient in the conversation.

- **Dictation**: In this mode, the service assumes that there is a single person narrating in the audio. This option is useful for clinicians dictating at the end of a patient visit.

Amazon Transcribe Medical provides APIs that work for both synchronous batch workflows and asynchronous streaming workflows.

Streaming

In streaming transcription, the text output from the service is generated in real time by taking in a stream of audio. You can also use an audio file and generate a stream of text from it. The service supports HTTP/2 and WebSocket protocols using which real-time transcriptions can be generated. You can also set up the streaming transcription using the AWS SDK or go to the AWS console to try it out. The following screenshot shows the options available under the real-time streaming option for Transcribe Medical.

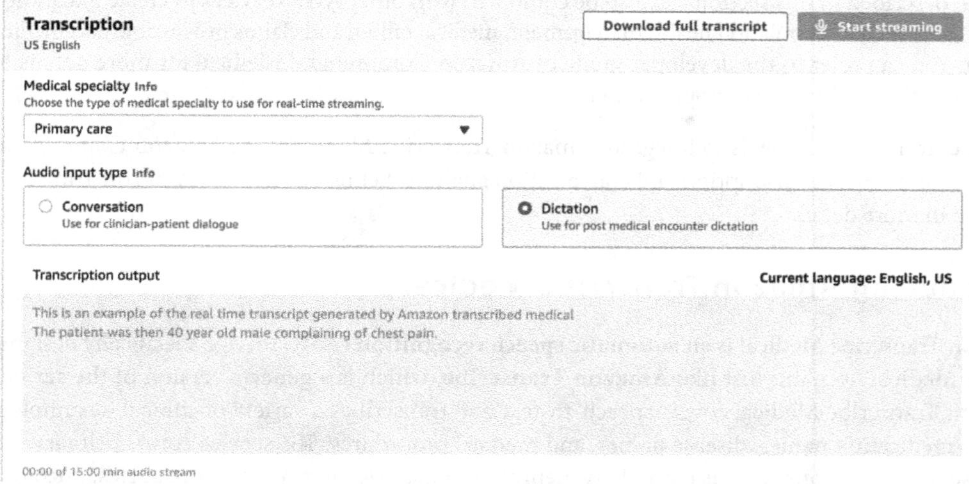

Figure 2.1 – The AWS console for Amazon Transcribe Medical real-time transcription

As shown in *Figure 2.1*, the service provides you with the ability to choose your medical specialty. It also provides options for choosing between conversation and dictation. The service supports the US English language and also recognizes different accents spoken by non-native English speakers. The output is generated in JSON format and supports word-level timestamps and confidence scores.

Batch

You can use Transcribe Medical to start a batch transcription job. You provide the location of the audio file that you want to transcribe. It supports multiple audio file formats such as FLAC and WAV. Once you start a transcription job, you can track its status and progress. The results of all batch transcription jobs are stored on S3, the object store from AWS. It provides the output in JSON format, which makes it easy to integrate with other AWS AI services and also within downstream applications. Let us look at the various options that Transcribe Medical provides for batch transcription in the AWS console.

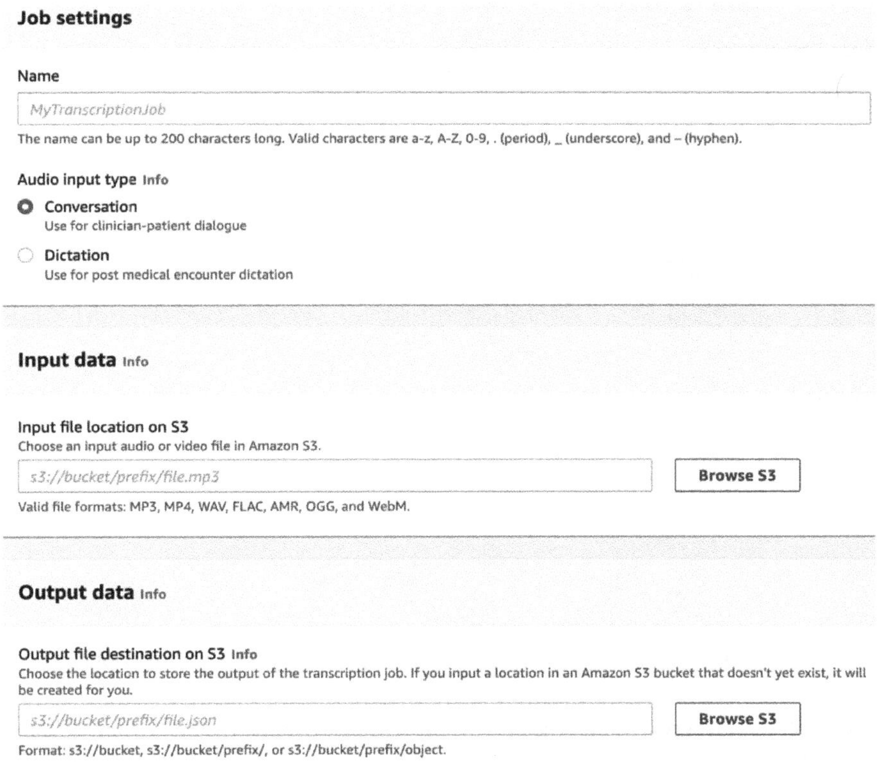

Job settings

Name

MyTranscriptionJob

The name can be up to 200 characters long. Valid characters are a-z, A-Z, 0-9, . (period), _ (underscore), and – (hyphen).

Audio input type Info
○ Conversation
   Use for clinician-patient dialogue

○ Dictation
   Use for post medical encounter dictation

Input data Info

Input file location on S3
Choose an input audio or video file in Amazon S3.

s3://bucket/prefix/file.mp3   **Browse S3**

Valid file formats: MP3, MP4, WAV, FLAC, AMR, OGG, and WebM.

Output data Info

Output file destination on S3 Info
Choose the location to store the output of the transcription job. If you input a location in an Amazon S3 bucket that doesn't yet exist, it will be created for you.

s3://bucket/prefix/file.json   **Browse S3**

Format: s3://bucket, s3://bucket/prefix/, or s3://bucket/prefix/object.

Figure 2.2 – The AWS console for Amazon Transcribe Medical batch transcription

As shown in the screenshot, you can select between conversation and dictation audio format. You can then provide the S3 location of your audio file that you want to transcribe and the output location where Transcribe Medical can store the output JSON file.

Custom vocabulary

Custom vocabulary allows you to improve the accuracy of Transcribe Medical output. This is useful for terms that the default Transcribe Medical models are not able to accurately convert to text from its pronunciation in the audio. It is also useful for synonyms or terminology that is specific to an organization that may be present in the audio. These help you to design a custom version of Transcribe Medical that is unique to your organization's use case.

Custom vocabulary is created using user input that is provided to Transcribe Medical via a tab-delimited text file. The text file provides the columns that the user needs to fill. These columns include the original phrase, how it should sound, and how it should be displayed. This provides the users an intuitive way to provide vocabulary to the service. Once the text file is created, it can be used to create a custom vocabulary in Transcribe Medical. The following screenshot shows the **Create vocabulary** screen that can be used in the AWS Management Console to create a custom vocabulary in Transcribe Medical.

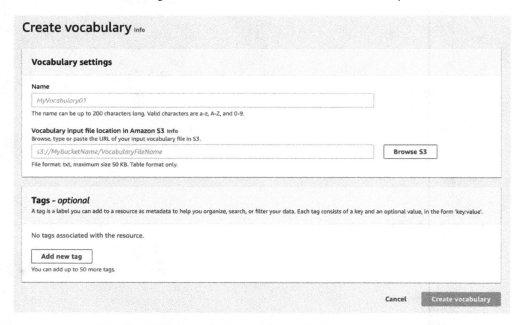

Figure 2.3 – The AWS console for Amazon Transcribe Medical custom vocabulary

As shown in the preceding screenshot, the custom vocabulary creation process requires the location of the input text file created by you. You can upload the text file on S3 and provide its location for Transcribe Medical to automatically create a custom vocabulary. Once the custom vocabulary is created, it can be used in both the real-time streaming transcriptions and batch transcription jobs by choosing the custom vocabulary option.

Amazon Transcribe Medical makes it simple for users to create workflows that involve voice-enabled inputs. Once the audio is transcribed, it can be fed into Amazon Comprehend Medical to detect key medical entities for analysis. You can refer to the developer guide of Amazon Transcribe Medical for a deeper understanding of the service and the APIs it provides.

Let us now look at Amazon HealthLake and how it helps you store, transform, and query healthcare data while adhering to **healthcare interoperability** standards using **Health Language International (HL7) Fast Healthcare Interoperability Resources (FHIR)**.

Introducing Amazon HealthLake

Amazon HealthLake is the latest addition to healthcare-specific AI services provided by AWS. It is a HIPAA-eligible service that allows you to ingest healthcare data in FHIR format. It then stores that data in a fully managed FHIR data store. During the ingestion process, it transforms the data by automatically extracting key medical entities from unstructured records in your healthcare data, thereby giving you a complete view of a patient's record. You can then query your records using both the structured portion of your patient records and also the entities extracted from the unstructured portions of your healthcare records.

Before we look into the details of HealthLake, let us understand the FHIR format better. FHIR is a healthcare interoperability standard created by HL7 with the aim of standardizing how healthcare information is represented and exchanged. It uses the JSON format to represent healthcare entities known as FHIR resources. The FHIR resource definition contains attributes that fully describe the healthcare entity. These FHIR resources are stored in a FHIR data store and can be connected together through natural keys in a query to get a full view of the patient. FHIR released multiple versions of FHIR standards and HealthLake supports the R4 version of FHIR. You can find more information about the FHIR R4 standard in their resource guide.

To begin using HealthLake, you will need FHIR R4-compliant data. This is a common healthcare interoperability standard and a large number of healthcare systems support FHIR-enabled interfaces that can provide FHIR resources to ingest into a HealthLake data store. There are AWS partners who can help if you do not have FHIR R4-compliant data. Once you have data in FHIR R4 format, you can create a data store and ingest data into it.

Creating and importing data into a data store

HealthLake allows you to create fully managed FHIR data stores. The data store is the repository where you store your FHIR resources. You can create the data store using the AWS CLI, the console, or APIs available via the AWS SDK. Here is an example of creating the FHIR data store in HealthLake using the AWS console.

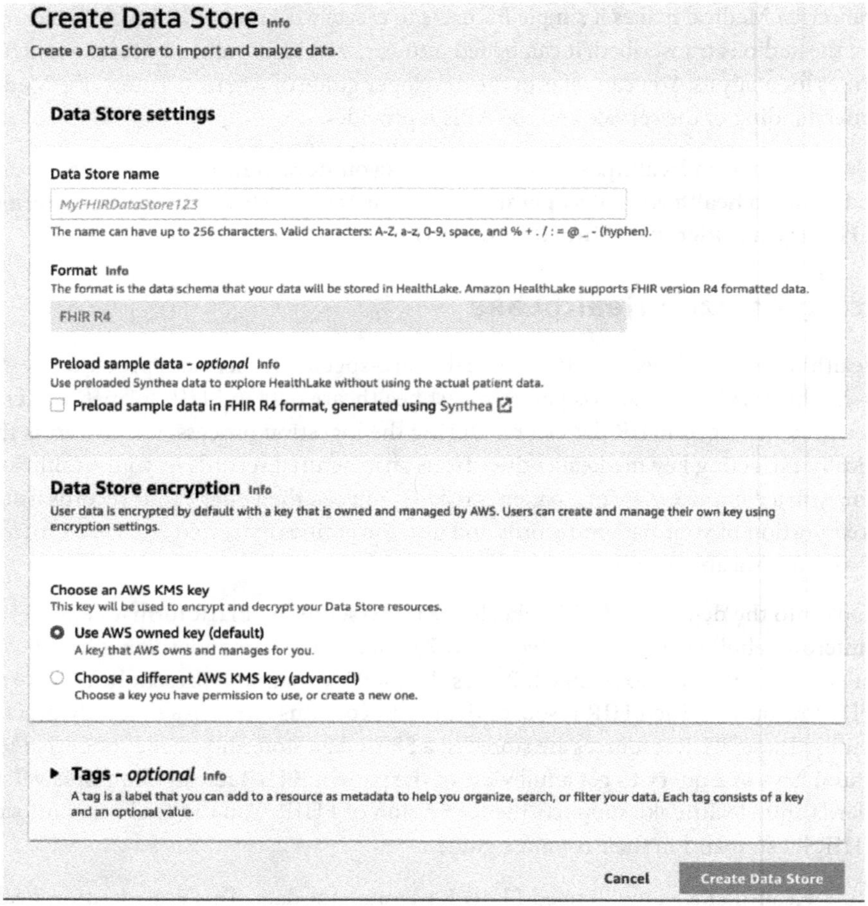

Figure 2.4 – The AWS console for Amazon HealthLake Create Data Store

As shown in the preceding screenshot, you can provide the name of your data store while creating it. During the creating process, you can choose to preload the data store with some artificially generated patient data using **Synthea**, which is a synthetic patient data generator. This allows you to experiment with the service without loading it with real-world patient data. The service also supports encryption using **AWS Key Management Service (KMS)** keys.

Once the data store has been created, you can import data into it using the AWS console, CLI, or API available via the AWS SDK. You can also interact with the records in the data store using a **Hypertext Transfer Protocol (HTTP)** client using methods such as POST, PUT, and GET.

If there are unstructured records in your medical data, Amazon HealthLake automatically detects clinical entities in it. The unstructured information is stored in the DocumentReference resource in FHIR. These entities are added as extensions in your DocumentReference FHIR resource and are available for search and query operations downstream. This is an automatic behind-the-scenes operation that HealthLake carries out on your behalf. The extensions are immediately available for search and query after ingesting into HealthLake.

Querying and exporting data from Amazon HealthLake

Amazon HealthLake data store supports FHIR-based search or query operations. You can also search using the GET or POST HTTP methods. It is highly recommended to use the POST operation to search when the attributes include PHI information. You can use the GET operation to search from the Management Console. Here is an example of performing the query operation from the AWS Management Console.

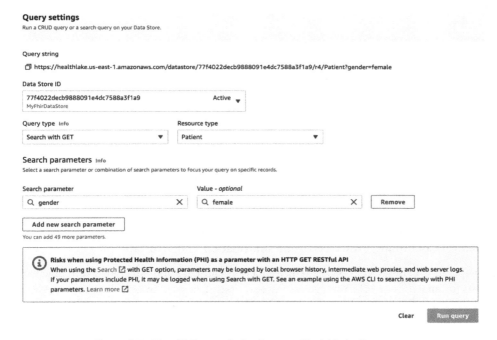

Figure 2.5 – The AWS console for Amazon HealthLake Run query

As you can see in the screenshot, the HealthLake query feature allows you to select the **Data Store ID**, **Query type**, **Resource type**, and the **Search parameters** on which you want to query. In this case, I have chosen to use the **GET** operation by selecting **Search with GET** to fetch all female patients in my data store. The results of this query are fetched from the underlying patient resource. As mentioned earlier, I am using synthetic patient data in this query that is generated by Synthea. If I had real-world patient data, I would have used the **POST** operation. As described earlier, this query functionality automatically extends into the unstructured part of your clinical data in HealthLake, which is stored in the **DocumentReference** resource. Here is an example of such a query from the AWS Management Console.

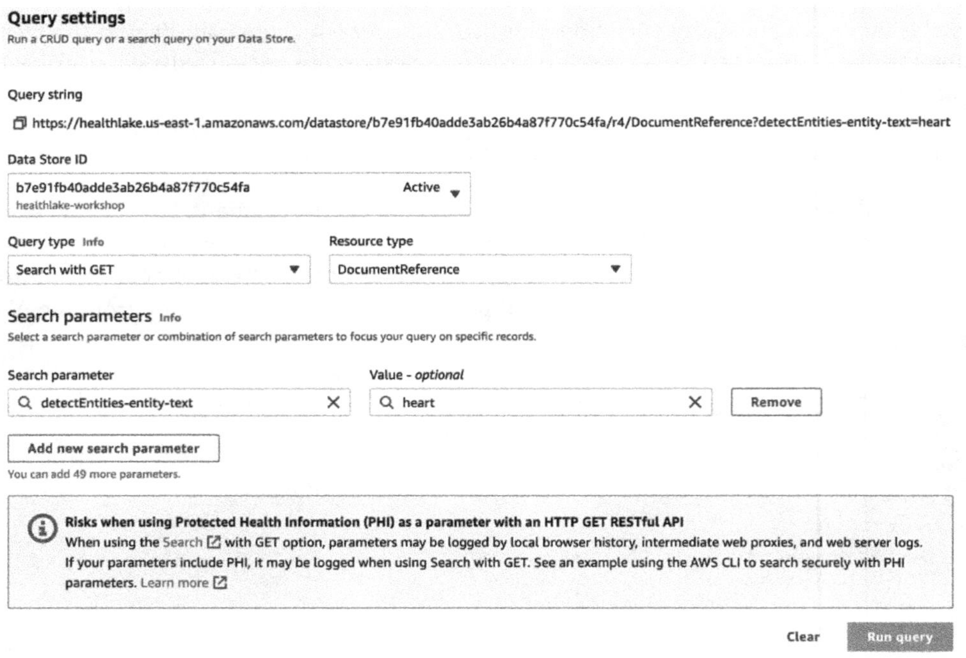

Query settings

Run a CRUD query or a search query on your Data Store.

Query string

🗇 https://healthlake.us-east-1.amazonaws.com/datastore/b7e91fb40adde3ab26b4a87f770c54fa/r4/DocumentReference?detectEntities-entity-text=heart

Data Store ID

| b7e91fb40adde3ab26b4a87f770c54fa | Active ▾ |
| healthlake-workshop | |

| Query type Info | Resource type |
| Search with GET ▾ | DocumentReference ▾ |

Search parameters Info

Select a search parameter or combination of search parameters to focus your query on specific records.

| Search parameter | Value - optional | |
| 🔍 detectEntities-entity-text   ✕ | 🔍 heart   ✕ | Remove |

| Add new search parameter |

You can add 49 more parameters.

ⓘ **Risks when using Protected Health Information (PHI) as a parameter with an HTTP GET RESTful API**
When using the Search 🗗 with GET option, parameters may be logged by local browser history, intermediate web proxies, and web server logs.
If your parameters include PHI, it may be logged when using Search with GET. See an example using the AWS CLI to search securely with PHI
parameters. Learn more 🗗

Clear   Run query

Figure 2.6 – The AWS console for Amazon HealthLake Run query with DocumentReference resource

As shown in the preceding figure, I can search the **DocumentReference** FHIR resource and select the detectEntities-entity-text parameter. I then provide the value heart for my query. This will allow me to filter for clinical notes that mention "heart." I can also select other parameters such as **Medical Conditions**, **Protected Health Information**, and **Medications**.

The query operation allows you to slice and dice data in HealthLake and integrate it with downstream applications for analytics or ML. You can also export your data store into S3. The export API from HealthLake exports data in **newline delimited JSON** (**ndjson**) format. Each line in the export file consists of a valid FHIR resource. You can perform the export operation from the CLI, AWS Console, or the API provided via the AWS SDK. For more details about Amazon HealthLake and its APIs, you can refer to the Developer Guide of Amazon HealthLake.

Amazon Comprehend Medical, Amazon Transcribe Medical, and Amazon HealthLake are three AI services from AWS that are specific to healthcare and life sciences. However, there are some other services from AWS that are highly utilized in healthcare and life sciences use cases even though they are not designed specifically for the domain. Next, let us look at some of these services.

Other AI services for healthcare and life sciences

As we saw in *Chapter 1*, the AWS AI/ML stack provides a wide array of services for ML use cases. Here are some key services from the stack that are useful for healthcare and life sciences use cases – we will be utilizing some of them in the subsequent chapters:

- **Amazon Textract**: Amazon Textract is an ML service from AWS that automatically extracts text and handwritten data from scanned documents. It automates the process of extracting and understanding data in PDFs and images. With Amazon Textract APIs, you can detect text in the document, key-value pairs in a form, and also detect the presence of a table and extract individual cell values from it. You can also ask questions to Textract about data in the document and Textract will provide a value from the document corresponding to the question. For example, you can analyze the patient intake form using Amazon Textract and ask *what is the patient's name?* and Textract will provide a response with the name of the patient.

  Amazon Textract provides both asynchronous APIs and synchronous APIs that you can integrate with your downstream applications. To learn more about Amazon Textract, you can refer to its developer guide.

- **Amazon SageMaker**: Amazon SageMaker allows data scientists to build, train, and deploy ML models at scale. While the AI services provide pre-trained models in most cases and require minimal knowledge of ML, SageMaker provides tools for data scientists to customize their ML models as needed. SageMaker is a feature-rich ML development environment and provides tools to label and prepare your data, build your models, train and tune your models, and deploy and manage your models. Each of these areas comes with multiple tools and options that you can choose from depending on your use case and the type of data.

  SageMaker also provides an **Integrated Development Environment (IDE)** for ML called **SageMaker Studio**. The IDE provides an easy-to-use interface for accessing all the features of SageMaker and also provides Jupyter notebooks for writing ML scripts. You also have the option of running AutoML on SageMaker using AutoPilot. For citizen data scientists, **SageMaker Canvas** provides an easy-to-use intuitive UI that takes you through the process of creating high-quality ML models. For more about Amazon SageMaker, you can refer to its developer guide.

Now that we have a high-level understanding of the key AI/ML services from AWS, you can start thinking about how to map them to particular use cases and address specific capabilities. Then, you can dive deeper into that service by referring to the corresponding developer guide. For example, you can check whether your use case involves any unstructured clinical text that can be processed using Amazon Comprehend Medical, whether you have to process and store FHIR format data that Amazon HealthLake can be utilized for, or whether you have to process and analyze multiple documents as part of your use case that Amazon Textract can help with. These building blocks of capabilities are unique in their own way, but when used together can create powerful intelligent applications. We will see examples of such use cases in future chapters of this book.

Summary

In this chapter, we went through the different segments of the healthcare and life sciences industry. These include healthcare providers, healthcare payors, pharmaceutical organizations, medical devices, and genomics. In each of these segments, we went over some important applications of ML. We then got an introduction to some key AI services from AWS specific to healthcare and life sciences, such as Amazon Comprehend Medical, Amazon HealthLake, and Amazon Transcribe Medical. We also touched upon other services such as Amazon Textract and SageMaker that are not specifically designed for the healthcare and life sciences domain but address specific problems in the industry.

In *Chapter 3, Machine Learning for Patient Risk Stratification*, we will learn about how ML can help identify at-risk patients for various types of clinical conditions or medical events. Timely identification of these patients helps avoid negative healthcare outcomes and improves the overall quality of care.

Further reading

- *Magic Quadrant for Cloud Infrastructure and Platform Services* by Raj Bala, Bob Gill, Dennis Smith, David Wright, Kevin Ji in Gartner (2021): `https://www.gartner.com/doc/reprints?id=1-271OE4VR&ct=210802&st=sb`

- *Solution Scorecard for Amazon SageMaker* by Daniel Cota, Su En Goh in Gartner (2021): `https://www.gartner.com/en/documents/4002618`

- *AWS Healthcare & Life Sciences Case Studies*: `https://aws.amazon.com/health/case-studies/?nc=sn&loc=5&refid=ar_card&case-studies-health-cards.sort-by=item.additionalFields.publishedDate&case-studies-health-cards.sort-order=desc&awsf.case-studies-filter-area=*all`

- *UK Biobank*: `https://www.ukbiobank.ac.uk/`

- *Centers for Medicare & Medicaid Services*: `https://www.cms.gov/medicare/icd-10/2021-icd-10-cm`

- *RxNorm Files*: `https://www.nlm.nih.gov/research/umls/rxnorm/docs/rxnormfiles.html`

- *SNOMED*: `https://www.snomed.org/snomed-ct/why-snomed-ct`

- *Amazon Comprehend Medical*: `https://docs.aws.amazon.com/comprehend-medical/latest/dev/comprehendmedical-welcome.html`

- *Amazon Transcribe Medical*: `https://docs.aws.amazon.com/transcribe/latest/dg/transcribe-medical.html`

- *FHIR resources*: `https://www.hl7.org/fhir/resourcelist.html`

- *Synthea*: `https://synthetichealth.github.io/synthea/`

- *Amazon HealthLake*: `https://docs.aws.amazon.com/healthlake/latest/devguide/what-is-amazon-health-lake.html`

- *Amazon Textract*: `https://docs.aws.amazon.com/textract/latest/dg/what-is.html`

Part 2: Machine Learning Applications in the Healthcare Industry

The next four chapters will dive deeper into some specific use cases for AWS in the healthcare sphere, including assessing patient risk and improving efficiency in various parts of the industry.

- *Chapter 3, Machine Learning for Patient Risk Stratification*
- *Chapter 4, Using Machine Learning to Improve Operational Efficiency for Healthcare Providers*
- *Chapter 5, Implementing Machine Learning for Healthcare Payors*
- *Chapter 6, Implementing Machine Learning for Medical Devices and Radiology Images*

3

Machine Learning for Patient Risk Stratification

Chapters 1 and *2* were foundational chapters that were designed to help you get introduced to the concepts of ML, its applications in healthcare and life sciences, and also some key services from AWS. With this foundational knowledge, we can now start applying these techniques to real-world industry problems. Over the course of the next eight chapters, you will learn about specific applications of AI/ML for solving problems for healthcare providers, payers, pharma, medical devices, and genomics customers.

In this chapter, we will look at one of the most common usages of ML in healthcare, the stratification or identification of risky patients. We will learn what it takes to identify risky patients and the common ML models that can help with this identification. We will then implement an example ML model that identifies patients at risk of breast cancer using SageMaker Canvas, the low-/no-code service from AWS that allows citizen data scientists to automatically build ML models using an intuitive user interface. These topics will be covered in these sections:

- Understanding risk stratification

- Implementing ML for patient risk stratification

- Introducing SageMaker Canvas

- Implementing ML for breast cancer risk prediction

Technical requirements

The following are the technical requirements that you need to have in place before building the example implementation at the end of this chapter:

- Set up Amazon SageMaker prerequisites. Follow the instructions on setting up the prerequisites to use Amazon SageMaker as described at the following link: `https://docs.aws.amazon.com/sagemaker/latest/dg/gs-set-up.html`.

- Complete the steps to set up SageMaker Canvas prerequisites. To achieve this, follow the instructions as described at the following link: `https://docs.aws.amazon.com/sagemaker/latest/dg/Canvas-getting-started.html#Canvas-prerequisites`. Make sure you are able to access the SageMaker Canvas console as indicated on the prerequisites page.

> **Note**
>
> You only need to complete the prerequisites section of SageMaker Canvas setup. Once you have access to the login screen, you do not need to complete the rest of the steps to complete the activity in the chapter.

Once you complete the above setup steps and have access to the SageMaker Canvas user interface, you can perform the steps described in the example implementation.

Understanding risk stratification

The goal of a well-functioning healthcare system is to uplift the health and well-being of the population with high-quality, timely care while keeping costs low. Healthcare providers are measured on their ability to do this by the **Centers for Medicare and Medicaid Services (CMS)** using a variety of metrics. The performance of healthcare providers against these metrics is an important factor in how they are reimbursed for the services they provide.

As part of the **value-based care model**, CMS introduced programs that reimbursed providers based on the value of care they provide to their patients. The traditional model, known as *fee-for-service*, incentivized providers in terms of the number of services they provided, such as the number of procedures they performed or the number of tests they ordered. This increased the overall cost of care without noticing a lot of improvements in the general health of the population. In the value-based care model, providers need to show how they are performing on the metrics created by CMS and publish that information for payers. These quality measures are created by taking into account their correlation to the overall improvement in the health of the population. Examples of such metrics include readmissions and mortality. The lower the number of readmissions and mortality among the patients treated by the provider, the better it is for them. Under the pay-for-performance initiative, providers are incentivized for better patient engagement, increased use of evidence-based medicine, and more reliance on data-driven analytical decision-making.

Identifying at-risk patients

A healthcare system needs to put a patient's health and well-being above everything else. However, not all patients need the same interventions and care. Some patients are really sick and may need increased attention and resources. If proper care and timely interventions are not provided, these patients risk having a negative outcome – which, in addition to the impact on the patient themselves, may adversely affect the provider's quality metric. It is therefore vital for the provider to identify the cohort of patients at risk of a negative medical outcome so extra attention can be given to them.

The method of identifying at-risk patients is driven by data collected during the patient's care journey, compared against the usual signs of risk. For example, a patient with high blood pressure and increased cholesterol levels is at risk of **congestive heart failure** (**CHF**). The measurement of blood pressure and cholesterol is usually done in any preventive health checkup and it's easy to identify these patients based on this data. However, there may be non-correlated attributes in the data that contribute to the negative medical event. These attributes are sometimes hidden in data formats that are not easy to process and extract information from. For example, signs of deteriorating health may be hidden in medical images or transcripts of doctor-patient exchanges. Also, medical information about a patient is derived over a period of time during the patient's care journey. Therefore, it may take a long time before all data points are available to classify a patient as risky. Sometimes, this extra time may be critical. This is because the earlier the providers identify risky patients, the more time they have to take meaningful actions to help the patient.

These underlying challenges cannot be addressed by conventional analytics and data processing techniques. As a result, providers are leaning towards ML-based approaches to identify risky patients early in the care journey. The risk identification technique varies depending on the disease or medical event in question and the available data. Let us look at some of the common approaches to implementing ML for patient risk stratification.

Implementing ML for patient risk stratification

ML algorithms can identify patterns in patient data that correlate with a medical event. For example, ML models can look at the observations, medical conditions, medications, and demographic information of chronic diabetes patients to determine whether they are at risk of kidney failure. As a result of this early identification, those patients can be provided additional intervention or medications that would prevent this from happening. There are several factors to keep in mind when choosing the right ML-based approach to stratify the patient population. These guidelines will help you define a problem and choose the right datasets and ML algorithms appropriate for the dataset and the problem. They will also help you determine whether the generation of a risk profile is useful or not based on how actionable the insights are. Here are a few important points to keep in mind:

- **Problem Definition and Dataset**: The first step to keep in mind is the definition of the problem of risk stratification from an ML perspective. Patient data can have multiple indicators that could result in a negative medical outcome. The goal of the ML model is to identify and correlate those key indicators to the negative outcome. In some cases, it's important to look at how the data has changed over a period of time. For example, how a patient's cholesterol levels have trended for the past 5 years can contribute a lot to their risk of congestive heart failure. In such cases, snapshots of data need to be gathered over a period of time to determine the trend. Depending on the disease, these trends can be tracked in the short term or long term. The available dataset along with the problem definition should set the stage for choosing the correct ML approach to be applied.

- **Choosing the right approach**: Once the problem has been defined, you need to choose the right ML approach to solve the problem. The approach is largely driven by our understanding of the data. Here are some common approaches and some pointers to help you select them:

  - **Clustering**: Clustering of patient records can help you group similar patients together. This approach is useful when we do not have available labels or in the absence of a target variable. By clustering groups of patients based on their medical records, we can identify key topics or dominant categories in each group that can help define a risk profile for them. For example, patients with high blood pressure and cholesterol may be all grouped together in one group whereas patients with high glucose and **HbA1C** values may be grouped into another group. The dominant category for the first group would be the risk of heart disease and for the second group could be the risk of type 2 diabetes.

  - **Time series analysis**: This approach is useful when we have disease progression data over a period of time mapped to an outcome. Time series-based approaches are extremely useful to track when a patient might be on a path to a negative outcome and prevent it from happening. You model the problem as a forecasting problem with a time attribute that has a set of observations at regular time intervals. For each of the observations, you have key disease metrics that you track. The model can then be trained to predict the likely outcome over a period of time for a patient who is on a similar path.

  - **Regression**: Regression models allow you to predict a number based on values of other features. When applied to risk stratification problems, it can help you predict a numerical risk score for a population. For example, the **Framingham risk score** (https://www.framinghamheartstudy.org/fhs-risk-functions/hard-coronary-heart-disease-10-year-risk/) is used to assess cardiovascular risk in patients based on features such as age, total cholesterol, and smoking status. You can use similar risk score assessments for other diseases that can be predicted using an ML regression model.

  - **Classification**: In this approach, you have a cohort of patients who are identified as having a certain disease or outcome and also another cohort who do not belong to that category. This forms the basis of a **binary classification model** that learns how to predict whether a new patient is positive or negative for that disease class. The model also generates a probability value for each class that can be used for risk assessment. We will be using this approach in our example later to predict the risk of breast cancer.

We can also combine approaches to create more complex models for risk stratification. For example, a dataset may contain observations for patient cohorts captured over a period of time and related to a target class of 1 (positive) or 0 (negative). You could potentially create classification models on such a dataset at different time intervals to see how the class probabilities change over a period of time. You can also associate a risk score that is calculated based on these results and track the risk score progression during a patient's life cycle.

- **Actionable**: The last thing to consider while using ML for risk stratification is to check whether the insights are actionable. There are two factors that can make the insight actionable. The timeliness of the insight is very important. Your model may be highly accurate in predicting the risk profile of the patient but if it is not able to do that with enough time for the provider to take meaningful action, it may not be useful.

Now that we have a good understanding of various ML methodologies for patient risk stratification, let us apply this learning to build a model using SageMaker Canvas. Let us go over SageMaker Canvas in more detail.

Introducing SageMaker Canvas

SageMaker Canvas is a no-code ML service that allows business analysts to create accurate ML models using a point-and-click interface. It supports the creation of regression, classification, and forecasting models that utilize tabular datasets in comma-delimited CSV format. The service runs from an internet browser and provides intuitive screens to perform various steps in building an ML model:

- **Importing and joining data**: SageMaker Canvas allows users to import data from a variety of sources on AWS such as Amazon S3 and Amazon Redshift. It also supports connecting and importing data from Snowflake. This provides a seamless way for users to start building their models on data stored in AWS or outside of AWS. In addition, the service also allows you to import data from your local computer. Once the dataset is imported into the service, you can view a sample of it and also join it with other imported datasets using a visual interface. Once you have imported and joined your data as needed, you can select it and move on to the next step of building your model.

- **Building your model**: In this step, the users pick a target variable to predict. Based on this selection, SageMaker Canvas automatically selects a model type to use for training. This can be a two-category (binary) classification model, a 3+ category classification model, a regression model, or a time series model. Users then have the option to choose features to include or exclude from the model-building process. For example, your dataset might have an ID column that you can exclude from the model-building process as it doesn't add any value to the model during training. SageMaker Canvas also handles missing values by calculating approximate missing values based on existing values. Next, you can go ahead and select **Quick Build**, which takes anywhere between 2-15 mins to build the model. Or, you can select **Standard Build**, which takes 2-4 hours and generally has better accuracy. Before building the model, you can choose to **preview** a model to get an idea of how the data in each column is distributed and also the estimated accuracy of the prediction.

- **Evaluating your model**: Once you have built your model, SageMaker Canvas allows you to evaluate your model performance by generating a few metrics automatically. For example, it generates a column impact percentage that allows you to determine how important a column was in making the prediction that the model made. It also provides a scoring section that allows you to determine model performance beyond the standard accuracy metric. For example, in the case of classification models, you can see the confusion matrix and the **F1 score** for the model and in the case of regression, you can see the **root mean square error** (**RMSE**) for the model.

- **Making predictions**: After you have built and evaluated your model, you can use it to make predictions on unseen data. SageMaker Canvas allows you to generate predictions as a batch or generate a single prediction from your model. If you choose batch prediction, SageMaker Canvas allows you to choose a CSV file with a missing target value to generate the predicted values. It then generates the predicted values and the probabilities of the predicted values for each row in the batch dataset. Once the predictions have been generated, you can download them as a CSV file. If you choose a single prediction, SageMaker Canvas provides you with a form to input the values of each feature in a row. It then uses those values to generate the predicted value of the target variable.

In addition to these core functionalities of importing data, joining datasets, building the model, evaluating the model, and generating predictions, SageMaker Canvas also provides other useful features for updating and sharing your models. For example, you can use the **update model** feature of SageMaker Canvas to retrain the models you have already built. This might be needed when new data is available or if the model performance is not up to the mark. It also versions each of the models so you can always go back to the previous model version if needed and delete them if they are obsolete. You can also use the model sharing feature of SageMaker Canvas to share your model with other data scientists. This is also useful when you want to deploy the model as an endpoint for integration with other applications.

> **Note**
>
> Model sharing is only possible for models that are built using Standard Build and for any other model type except time series models. For more details about SageMaker Canvas, you can refer to the section on SageMaker Canvas in the SageMaker Developer Guide (`https://docs.aws.amazon.com/sagemaker/latest/dg/Canvas.html`).

Now that we have an understanding of SageMaker Canvas, let us use it to build a model for risk stratification for breast cancer patients. We will go through the steps to import, build, analyze, and generate predictions from the model using this example implementation.

Implementing ML for breast cancer risk prediction

Breast cancer affects about 1 in 8 women in the US. In 2020, more than 2.3 million women were diagnosed with breast cancer worldwide and 685,000 died as a result. These grim statistics provide enough information for us to conclude that breast cancer is a deadly disease. There have been several procedures used to diagnose breast cancer, from genomic testing to imaging-based studies. One common method is to look at the characteristics of the cell nuclei derived from imaging studies and classify them as malignant (M) or benign (B). In this example implementation, we will use this method to predict whether a breast mass is M or B using cell nuclei features. This prediction can be generated at various stages of the progression of the disease as the features of the cell nuclei change. This will help us determine whether a patient is at risk of developing a malignant breast mass over a period of time. Early determination of this risk and timely intervention can help prevent this negative outcome.

The dataset used in this example implementation comes from the UCI Machine Learning Repository (`https://archive.ics.uci.edu/ml/datasets/breast+cancer+wisconsin+(diagnostic)`). It consists of cell nuclei features such as radius, texture, perimeter, and area. There are a total of 30 such features in the dataset with 569 rows. One feature has the target label "M" or "B" denoting whether the breast mass is malignant or benign. We will build a model to predict this value using SageMaker Canvas. Before you begin, make sure you have set up SageMaker Canvas by following the steps highlighted in the getting started guide (`https://docs.aws.amazon.com/sagemaker/latest/dg/Canvas-getting-started.html`).

Importing your dataset

Let's walk through it:

1. You can download the dataset for this implementation from Kaggle (`https://www.kaggle.com/datasets/uciml/breast-cancer-wisconsin-data`). When you launch SageMaker Canvas for the first time, you will see an option to create a new model as shown in the following figure.

Figure 3.1 – The welcome screen when you launch SageMaker Canvas for the first time

2. Click on **+ New model** to create a new model. Provide a model name of your choice on the next screen. For example, `Breast Cancer Model`. Or, you can leave the default name provided by SageMaker Canvas. Then, click on **Create**.

    Next, you will see a screen with an option to **Import data to Canvas**, as shown in the following figure.

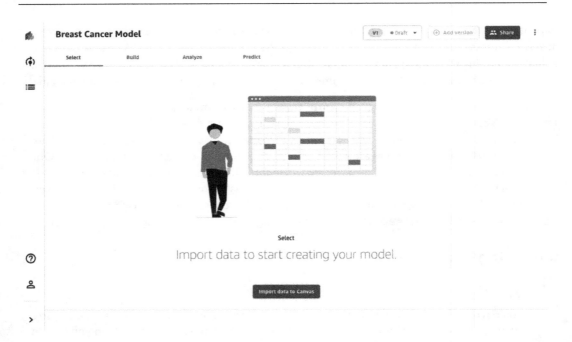

Figure 3.2 – Import data screen in SageMaker Canvas

3. Click on the button to be taken to the import data screen. Click on the **Upload** option to upload the data file you got from Kaggle earlier to SageMaker Canvas. Select the file and click **Open**. Then click **Import data**.

> **Note**
>
> If you see a message saying **Local file upload needs to be set up**, follow the instructions provided at the **Learn More** link to update the permissions on the S3 bucket.

Once the dataset is uploaded, you will see a status of **Ready** on the following, **Select dataset** screen, as shown in the following figure. You can also preview the dataset and examine a sample of records to see how it looks.

Figure 3.3 – Select dataset screen for SageMaker Canvas

4. Click on **Select dataset** to move on to the build phase.

Building the model

Next, we will use the data imported into SageMaker Canvas to build a model:

1. The first step you will need to do is select the target variable. In our case, the target variable is the column diagnosis that has the values of M or B. Select that column from the dropdown as shown in the following screenshot.

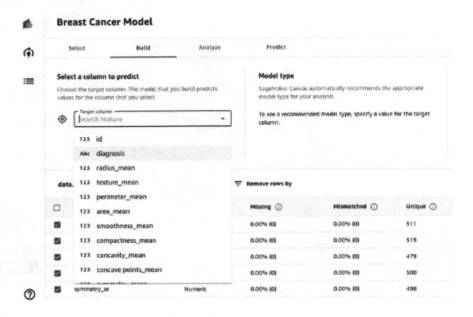

Figure 3.4 – Target variable dropdown on SageMaker Canvas model building screen

2. After selecting the target variable as **diagnosis**, SageMaker Canvas automatically determines the model type as **2 category prediction** or binary prediction. It also provides tabular statistics about the columns in the dataset such as the number of unique values, the mean/mode of the column values, and the correlation to the target variable. At this stage, you can choose to include all columns from the dataset or deselect some columns if they don't add value to the overall model. In our example, we have an **id** column that we can unselect as it just consists of a unique identifier for each patient. Also, **column _c32** has no values in it and is automatically deselected.

Figure 3.5 – Column statistics and model type on SageMaker Canvas model building screen

3. Next, you can preview the model to generate an *estimated accuracy* and *feature importance*. The estimated accuracy gives you an idea of how good the model is and looks at the important features that contribute to the model's performance. You can decide at this stage whether you would like to move forward with the build of the model or go back and change the dataset or features. Once you are ready to move forward, click on **Quick build** to build the model for analysis and prediction and wait for the build process to complete.

Analyzing the model

In this step, we will examine the model that we have built in more detail by looking at the column impact on the predicted value and the model scoring by examining its metrics:

1. The following figure shows the model analysis screen once our model is built.

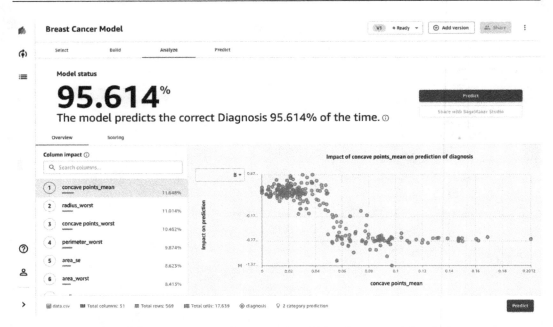

Figure 3.6 – Column impact visualization on SageMaker Canvas model analysis screen

As shown in the figure, you get a model accuracy score at the top of the model analysis screen that shows the accuracy of our model. Our model is over 95% accurate. In addition, for each column in our dataset, the screen provides a value of feature impact and a scatter plot of how the values of the feature compare with respect to its impact in predicting the B or the M value of the target variable. You can scroll through the list of columns and examine the impact score and the scatter plot for each column.

2. Next, in the **Scoring** tab of the analysis screen, you can see the model accuracy insights as shown in the following figure.

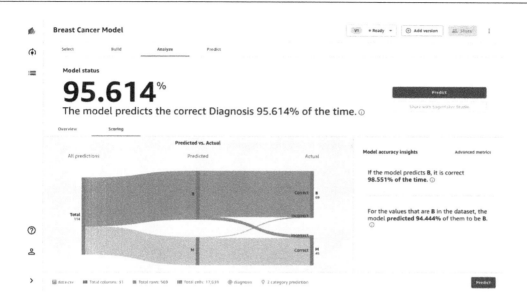

Figure 3.7 – Scoring tab of the SageMaker Canvas model analysis screen

As shown in the figure, the **Scoring** tab provides details of how the model performs for predicting true positive and true negative values. In our case, the model does a good job in predicting the B class where it is correct over 98% of the time. You can also look at advanced metrics for the model by clicking on **Advanced metrics**.

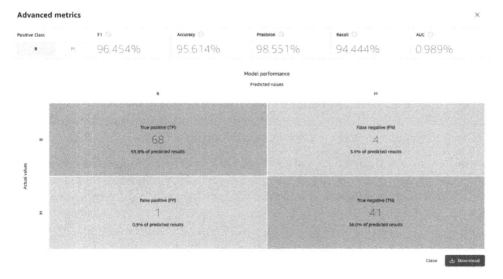

Figure 3.8 – Advanced metrics on the SageMaker Canvas model analysis screen

3. The **Advanced metrics** screen provides additional evaluation metrics for our model. For our binary classification problem, it provides a confusion matrix for both the B and the M class. It also provides **F1 score**, **precision**, **recall**, and **AUC**. As you can see, the model is performing well and we are now ready to use this model for prediction.

Predicting from the model

We are now ready to generate predictions from the model:

1. To begin predicting, click on **Predict** at the bottom of the model analysis screen. You can generate a batch prediction from the model by selecting a dataset with missing target variables.

2. In addition, you have the option to generate a single prediction for a set of values provided in real time. To generate a single prediction, select the **Single prediction** tab as shown in the following figure.

Figure 3.9 – Single prediction tab on the SageMaker Canvas model prediction screen

As shown in *Figure 3.9*, you can select different values for the input features on the left and generate the corresponding value of the prediction on the right. The output shows the prediction value (B, M) and also a probability value for each class. As mentioned at the beginning of the section, this probability value can change over a period of time, providing important insight into disease progression. This model can be run every time there are new values available to generate a new risk profile for the patient.

Now that you have gone through the steps of building an ML model using SageMaker Canvas, you can experiment with its other capabilities, such as building a model using the **Standard Build** option, which allows you to export and share the model with other data scientists and also deploy it as an endpoint.

Summary

In this chapter, we got an understanding of what patient risk stratification is and why it is important. We also got insights into why conventional analytical approaches may not be enough to stratify patients for risk. We then looked at various steps and guidelines before embarking on building a patient risk stratification model.

We got introduced to SageMaker Canvas, the no-code service from AWS that allows business analysts to build ML models.

Lastly, we went through an exercise creating an ML model to identify whether a breast mass is malignant or benign based on cell nuclei features and learned how this could help prevent the disease from taking a turn for the worse.

In *Chapter 4, Using Machine Learning to Improve Operational Efficiency for Healthcare Providers*, we will learn about how ML can help make healthcare providers more efficient in providing patient care by automating certain time-consuming tasks.

4
Using Machine Learning to Improve Operational Efficiency for Healthcare Providers

Operating an efficient healthcare system is a challenging but highly rewarding endeavor. As discussed in *Chapter 3*, healthcare systems are recognized for better care quality and saving costs. An efficient healthcare system innovates and focuses its time on improving patient care while reducing the time spent on repeatable operational tasks. It allows providers to focus on their core competency – taking care of patients – and reduce distractions that prevent them from doing this.

Patients in many places today have choices between multiple healthcare providers. The more patients a hospital is able to treat, the better it is so it can meet its revenue goals and remain profitable. However, a high volume of patients doesn't always equate to a high quality of care, so it's critical to strike a balance.

There are multiple areas of optimization in a healthcare workflow. Some of them are straightforward and can be implemented easily. Others require the use of advanced technologies such as **machine learning (ML)**.

In this chapter, we will dive deeper into the topic of operational efficiency in healthcare by looking at some examples and the impact that increased efficiency can have on hospital systems. We will then dive deeper into two common applications of ML for reducing operational overhead in healthcare – clinical document processing and dealing with voice-based applications. Both these areas require substantial manual intervention and the use of ML can help automate certain repeatable tasks. We will then build an example application for smart medical transcription analysis using the AWS AI and ML services. The application transcribes a sample audio file containing a clinical dictation and extracts key clinical entities from the transcript for clinical interpretation of the transcript and further downstream processing.

The chapter is divided into the following sections:

- Introducing operational efficiency in healthcare
- Automating clinical document processing in healthcare
- Working with voice-based applications in healthcare
- Building a smart medical transcription analysis application on AWS

Technical requirements

The following are the technical requirements that you need to complete before building the example implementation at the end of this chapter:

- Set up an AWS account and create an administrator user, as described in the Amazon Transcribe Medical getting started guide here: `https://docs.aws.amazon.com/transcribe/latest/dg/setting-up-med.html`.
- Set up the AWS **Command Line Interface** (**CLI**) as described in the Amazon Comprehend Medical getting started guide here: `https://docs.aws.amazon.com/comprehend-medical/latest/dev/gettingstarted-awscli.html`.
- Install Python 3.7 or later. You can do that by navigating to the Python **Downloads** page here: `https://www.python.org/downloads/`.
- Install the Boto3 framework by navigating to the following link: `https://boto3.amazonaws.com/v1/documentation/api/latest/guide/quickstart.html`.
- Install Git by following the instructions at the following link: `https://github.com/git-guides/install-git`.

Once you have completed these steps, you should be all set to execute the steps in the example implementation in the last section of this chapter.

Introducing operational efficiency in healthcare

For a business to run efficiently, it needs to spend the majority of its time and energy on its core competency and reduce or optimize the time spent on supporting functions. A healthcare provider's core competency is patient care. The more time physicians spend with patients providing care, the more efficient or profitable a healthcare system is.

One of the primary areas to optimize in a healthcare system is resource utilization. Resources in a hospital could be assets such as medical equipment or beds. They could be special rooms such as **operating rooms** (**ORs**) or **intensive care units** (**ICUs**). They could be consumables such as gloves and syringes. Optimal usage of these resources helps hospitals reduce waste and save costs. For example, ORs are the rooms in a hospital that earn the highest revenue, and leaving them underutilized is detrimental to a hospital's revenue. Using standard scheduling practices may leave out important considerations, such as the type of case before and after a particular case. Having similar cases scheduled back to back can minimize the setup time for the second case. It will also allow surgeons to be appropriately placed so they spend less time moving between cases. All these aspects add up to increase utilization of an OR room.

Another area of optimization is hospital bed utilization. It is a very important metric that helps hospitals plan a number of things. For instance, a low utilization of beds in a certain ward on a particular day could mean the hospital staff in that ward can be routed to other wards instead. Moreover, if the hospital can accurately estimate the length of stay for a patient, they can proactively plan for staffing adequate support staff and consumable resources. It will reduce pressure on employees and help them perform better over extended periods of time.

The concept of appropriate resource utilization also extends to reducing wait times. As a patient moves through their care journey in a hospital, they may be routed to multiple facilities or their case may be transferred to different departments. For example, a patient may come for a regular consultation at a primary care facility. They may then be routed to a lab for sample collection and testing. The case may then be sent to the billing department to generate an invoice and process a payment. The routing and management of the case is much more complicated for inpatient admissions and specialty care. It is important that the case moves seamlessly across all these stages without any bottlenecks. Using appropriate scheduling and the utilization of the underlying resources, hospitals can ensure shorter wait times between the stages of patient care. It also improves patient experience, which is a key metric for hospitals to retain and attract new patients to their facilities. A long wait time in the ER or an outpatient facility results in a bad patient experience and reduces the chances of patient retention. Moreover, waiting for labs or the results of an X-ray can delay critical medical procedures that depend on receiving those scan results. These are sequential steps and a delay in one step delays the whole operation.

The reduction of wait times requires the identification of steps in each area that are bottlenecks, finding ways to minimize those bottlenecks and accelerate the processes as a result. One of the common reasons for these kinds of delays is the result of manual intervention: depending on humans for repeatable tasks that can be automated reduces productivity. Using ML-based automation can help improve throughput and reduce errors. In the next two sections, we will look at two example categories of automation in a healthcare system: document processing and working with voice-based solutions.

Automating clinical document processing in healthcare

Clinical documentation is a required byproduct of patient care and is prevalent in all aspects of care delivery. It is an important part of healthcare and was the sole medium of information exchange in the past. The digital revolution in healthcare has allowed hospitals to move away from paper documentation to digital documentation. However, there are a lot of inefficiencies in the way information is generated, recorded, and shared via clinical documents.

A majority of these inefficiencies lie in the manual processing of these documents. ML-driven automation can help reduce the burden of manual processing. Let us look at some common types of clinical documents and some details about them:

- **Discharge summaries**: A discharge summary summarizes the hospital admission for the patient. It includes the reason for admission, the tests, and a summary of next steps in their care journey.

- **Patient history forms**: This form includes historical medical information for the patient. It is updated regularly and is used for the initial triaging of a patient's condition. It also contains notes from past physicians or specialists to understand the patient's condition or treatment plan better.

- **Medical test reports**: Medical tests are key components of patient care and provide information for physicians and specialists to determine the best course of action for the patient. They range from regular tests such as blood and urine to imaging tests such as MRI or X-ray scans and also specialized tests such as genetic testing. The results of the tests are captured in test reports and shared with the patients and their care team.

- **Mental health reports**: Mental health reports capture behavioral aspects of the patients assessed by psychologists or licensed medical councilors.

- **Claim forms**: Subscribers to health plans for a health insurance company need to fill out a claim form to be reimbursed for the out-of-pocket expenses they incur as part of their medical treatment. It includes fields for identifying the care provided, along with identifiable information about the patient.

Each of these forms stores information that needs to be extracted and interpreted in a timely manner, which is largely manually done by people. A major side effect of this manual processing is *burnout*.

Healthcare practitioners spend hours in stressful environments, working at a fast pace to keep up with the demands of delivering high-quality healthcare services. Some studies estimate over 50% of clinicians experiencing burnout in recent years. High rates of physician burnout introduce substantial quality issues and adversely affect patients. For example, stressed physicians are more likely to stop practicing, which puts patients at risk of having gaps in their care. Moreover, clinicians are prone to errors in judgement when making critical decisions about a patient's care journey, which is a safety concern. Research has also shown burnout causes boredom and loss of interest on the part of clinicians providing care, which does not bode well for patients who are looking for personalized attention. This may lead to patients dropping out of regular visits due to bad experiences. One answer to preventing burnout is to help clinicians spend less of their time on administrative and back-office support work such as organizing documentation and spend more time in providing care. This can be achieved through the ML-driven automation of clinical document processing.

The automation of clinical document processing essentially consists of two broad steps – firstly, extracting information from different *modalities* of healthcare data. For example, healthcare information can be hidden in images, forms, voice transcripts, or handwritten notes. ML can help extract information from these images using deep learning techniques such as **image recognition**, **optical character recognition**

(**OCR**), and **speech-to-text**. Secondly, the information needs to be interpreted or understood for meaningful action or decisions based on it to be taken. The solution to this can be as simple as a rules engine that carries out the post-processing of extracted information based on human-defined rules. However, for a more complex, human-like interpretation of healthcare information, we need to rely on **natural language processing** (**NLP**) algorithms. For example, a rules engine can be used to find a keyword such as "*cancer*" in a medical transcript generated by a speech-to-text algorithm. However, we need to rely on an NLP algorithm to determine the sentiment (sad, happy, or neutral) in the transcript. The complexity involved in training, deploying, and maintaining deep learning models is a deterrent for automation. In the final section of this chapter, we will see how AWS AI services can help by providing pre-trained models that make it easy to use automation in medical transcription workflows. Before we do that, let's dive deeper into voice-based applications in healthcare.

Working with voice-based applications in healthcare

Voice plays a big role in healthcare with multiple conversational and dictation-based types of software relying on speech to capture information critical to patient care. Voice applications in healthcare refer to applications that feature speech as a central element in their workflows. These applications can be patient-facing or can be internal applications only accessible to physicians and nurses. The idea behind voice-based applications is to remove or reduce the need for hands-on keyboard so workflows can be optimized and interactions become more intuitive. Let us look at some common examples of voice-based applications in healthcare:

- **Medical transcription software**: This is by far the most commonly used voice-based application in healthcare. As the name suggests, medical transcription software allows clinicians to dictate medical terminologies and convert them into text for interpretation, and store them in an **Electronic Health Record** (**EHR**) system. For example, a physician can summarize a patient's regular visit by speaking into a dictation device that then converts their voice into text. By using advanced analytics and ML, medical transcription software is becoming smarter and is now able to detect a variety of clinical entities from a transcription for easy interpretation of clinical notes. We will see an example of this in the final section of this chapter, where we build a smart medical transcription solution. Using medical transcription software reduces the time required and errors in processing clinical information.

- **Virtual assistants and chatbots**: While medical transcription solutions are great for long-form clinical dictations, there are some workflows in healthcare that benefit from short-form voice exchanges. For example, the process of searching for a healthcare specialist or scheduling an appointment for a visit involves actions such as filtering the type of specialization, looking up their ratings, checking their availability, and booking an appointment. You can make these actions voice-enabled to allow users to use their voices to perform the search and scheduling. The application acts as a virtual assistant, listening to commands and performing actions accordingly. These assistants are usually available to the end users in a variety of forms. For example, they can be accessed via chatbots running on a web application. They can also be evoked using a voice command on a mobile device – for instance, Alexa. They can also be packaged into an edge device that can be sitting in your home (such as an Amazon Echo Dot) or a wearable (such as the Apple Watch). At the end of the day, the purpose of virtual assistants is to make regular tasks easy and intuitive through voice, and it is vital to choose the right mode of delivery so the service does not feel like a disruption in regular workflows. In fact, the more invisible the virtual assistant is, the better it is for the end user. Virtual assistants have been successfully implemented throughout the healthcare ecosystem, whether in robotic surgeries, primary care visits, or even bedside assistance for admitted patients.

- **Telemedicine and telehealth services**: Telehealth and telemedicine visits have become increasingly popular among patients and providers lately. For patients, they provide a convenient way to get access to healthcare consultation from the comfort of their home or anywhere they may be, as long as they have network connectivity. It avoids the travel and wait times often required by in-person visits. For providers, it is easier to consult with patients remotely, especially if the consultation is in connection with an infectious disease. No wonder telehealth became so popular during the peak surge of the Covid-19 pandemic in 2021. As you can imagine, voice interactions are a big part of telemedicine and telehealth. Unlike medical transcriptions, which involve a clinician dictating notes, telehealth involves conversations between multiple parties, such as doctors, patients, and nurses. Hence, the technology used to process and understand such interactions needs to be different from medical transcription software. A smart telehealth and telemedicine solution allows multiple parties in a healthcare visit to converse with each other naturally and convert each of their speech portions to text, assigning labels to each speaker so it's understandable who said what in the conversation. This technology is known as **speaker diarization** and is an advanced area of speech-to-text deep learning algorithms. Moreover, NLP algorithms that derive meaning from these interactions should be able to handle conversational data, which is different from deriving meaning from single-person dictations.

- **Remote patient monitoring and patient adherence solutions**: Remote patient monitoring and patient adherence solutions are a new category of voice-based applications that providers use to ensure patients are adhering to the steps in their care plan, which is an integral part of patient care. Whether it is the adherence to medications as prescribed or adherence to visits per a set schedule, it is vital that patients follow the plan to ensure the **healthcare continuum**, which essentially means continuous access to healthcare over an extended period of time. It is estimated that about 50% of patients do not take their medications as prescribed and providers have little control over this metric while the patients are at home. Hence, a remote patient monitoring solution allows you to monitor a patient's adherence to a care plan and also nudge them with proactive alerts when it's detected that the patient is veering off course. These categories of solutions also provide reminder notifications when it's time to take a medication or time to leave for an appointment. They can also send recommended readings with references to information about the medications or the clinical conditions that the patient may be experiencing. These actions and notifications can be voice-enabled to make them more natural and easier to interact with, keeping them less disruptive.

As you can see, verbal communication plays a critical role in healthcare, so it's important that it's captured and processed efficiently for healthcare providers to be able to operate a high-quality and operationally efficient healthcare organization. Let us now see how we can build a smart medical transcription application using AWS AI services.

Building a smart medical transcription application on AWS

Medical transcripts document a patient's encounter with a provider. The transcripts contain information about a patient's medical history, diagnosis, medications, and past medical procedures. They may also contain information about labs or tests and any referral information if applicable. As you may have guessed, processing and extracting medical transcripts is critical to getting a full understanding of a patient's condition. The medical transcripts are usually generated by healthcare providers who dictate notes that capture this information. Then, medical transcriptions are manually transcribed into notes by medical transcriptionists. These notes are stored in the EHR system to capture a patient's encounter summary. The manual transcription and understanding of medical information introduce operational challenges in a healthcare system, especially when done at scale. To solve this problem, we will build a smart medical transcription solution that automates the following two steps:

1. Automating the conversation of an audio transcript into a clinical note using Amazon Transcribe Medical.

2. Automating the extraction of clinical entities such as medical conditions, **protected health information** (**PHI**), and medical procedures from the clinical note so they can be stored in an EHR system.

Let us now look at the steps to set up and run this application.

Creating an S3 bucket

Please make sure you have completed the steps mentioned in the *Technical requirements* section before you attempt the following steps:

1. Navigate to the AWS console using the `console.aws.amazon.com` URL and log in with your IAM username and password.

2. In the search bar at the top, type `S3` and click on the S3 link to arrive at the S3 console home page.

3. Click on the **Create bucket** button.

4. On the **Create bucket** console page, under **General configuration**, provide the name of your bucket in the **Bucket name** textbox, as shown in *Figure 4.1*:

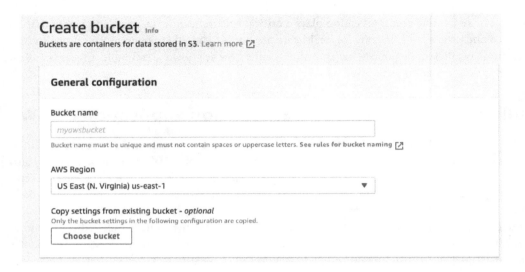

Figure 4.1 – The S3 Create bucket screen in the AWS Management Console

5.   Leave the rest of the choices as the default and click **Choose bucket** at the bottom of the screen.

This should create the S3 bucket for our application. This is where we will upload the audio file that will act as an input for Amazon Transcribe Medical.

Downloading the audio file and Python script

After you have created the S3 bucket, it's time to download the Python script and the audio file from GitHub:

1.   Open the terminal or command prompt on your computer. Clone the Applied-Machine-Learning-for-Healthcare-and-Life-Sciences-using-AWS repository by typing the following command:

```
Git clone https://github.com/PacktPublishing/Applied-
Machine-Learning-for-Healthcare-and-Life-Sciences-using-
AWS.git
```

You should now see a folder named Applied-Machine-Learning-for-Healthcare-and-Life-Sciences-using-AWS.

2.   Navigate to the code files for this exercise located at Applied-Machine-Learning-for-Healthcare-and-Life-Sciences-using-AWS/chapter-4/. You should see two files in the directory: audio.flac and transcribe_text.py.

3.   On the AWS console, navigate to the S3 bucket created in the previous section and click **Upload**.

4. On the next screen, select **Add files** and select the `audio.flac` audio file. Click **Upload** to upload the file to S3:

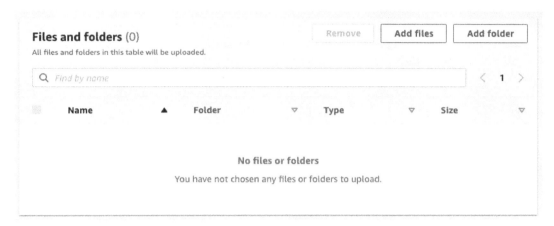

Figure 4.2 – The Add files screen of the S3 bucket in the AWS Management Console

5. Once uploaded, you should see the file in your S3 bucket. Click on it and open the `audio.flac` info screen. Once there, click **Copy S3 URI**:

Figure 4.3 – The object info screen in S3 with the Copy S3 URI button

6. Next, open the `transcribe_text.py` file on your computer. Scroll to line 10 and change the value of the `job_uri` variable by pasting the URI copied in the previous step between the double quotations. The S3 URI should be in the following format: `s3://`**bucket-name**`/key/audio.flac`. Also, input the S3 output bucket name on lines 25 and 37. You can have the same bucket name for input and output. Make sure to save the file before exiting.

You are now ready to run the Python script. You can look at the rest of the script to understand the code and read more about the APIs being used. You can also play the `audio.flac` audio file using an audio player on your computer to hear its contents.

Running the application

Now, we are ready to run this application. To run, simply execute the Python script, `transcribe_text.py`. Here are the steps:

1. Open the terminal or CLI on your computer and navigate to the directory where you have the `transcribe_text.py` file.

2. Run the script by typing the following:

```
python transcribe_text.py
```

The preceding script reads the audio file from S3, transcribes it using Transcribe Medical, and then calls Comprehend Medical to detect various clinical entities in the transcription. Here is how the output of the script looks in the terminal:

```
bash-4.2$ python transcribe_text.py
creating new transcript job med-transcription-job
Not ready yet...
Not ready yet...
Not ready yet...
Not ready yet...
Not ready yet...
Not ready yet...
Not ready yet...
Not ready yet...
Not ready yet...
Not ready yet...
Not ready yet...
transcription complete. Transcription Status:   COMPLETED
****Transcription Output***
The patient is an 86 year old female admitted for
evaluation of abdominal pain. The patient has colitis
and is undergoing treatment during the hospitalization.
The patient complained of shortness of breath, which is
worsening. The patient underwent an echocardiogram, which
shows large pleural effusion. This consultation is for
further evaluation in this regard.
Transcription analysis complete. Entities saved in
entities.csv
```

The preceding output shows the transcription output from Transcribe Medical. The script also saves the output of Transcribe Medical in the raw JSON format as a file named transcript.json. You can open and inspect the contents of the file. Lastly, the script saves the entities extracted from Comprehend Medical into a CSV file named entities.csv. You should see the following in the CSV file:

Id	Text	Category	Type
7	86	PROTECTED_HEALTH_INFORMATION	AGE
8	evaluation	TEST_TREATMENT_PROCEDURE	TEST_NAME
1	abdominal	ANATOMY	SYSTEM_ORGAN_SITE
3	pain	MEDICAL_CONDITION	DX_NAME
4	colitis	MEDICAL_CONDITION	DX_NAME
9	treatment	TEST_TREATMENT_PROCEDURE	TREATMENT_NAME
5	shortness of breath	MEDICAL_CONDITION	DX_NAME
10	echocardiogram	TEST_TREATMENT_PROCEDURE	TEST_NAME
2	pleural	ANATOMY	SYSTEM_ORGAN_SITE
6	effusion	MEDICAL_CONDITION	DX_NAME
12	evaluation	TEST_TREATMENT_PROCEDURE	TEST_NAME

Table 4.1 – The contents of the entities.csv file

These are the entities extracted from Comprehend Medical from the transcript. You can see the original text, the category, and the type of entity recognized by Comprehend Medical. As you can see, these entities are structured into rows and columns, which is much easier to process and interpret than a blurb of text. Moreover, you can use this CSV file as a source of your analytical dashboards or even ML models to build more intelligence into this application.

Now that we have seen how to use Amazon Transcribe Medical and Amazon Comprehend medical together, you can also think about combining other AWS AI services with Comprehend Medical for different types of automation use cases. For example, you can combine **Amazon Textract** and Comprehend Medical to automatically extract clinical entities from medical forms and build a smart document processing solution for healthcare and life sciences. You can combine the output of Comprehend Medical with **Kendra**, a managed search service from AWS to create a document search application driven by graph neural networks. All these smart solutions reduce manual intervention in healthcare processes, minimize potential errors, and make the overall health system more efficient.

Summary

In this chapter, we looked at the concepts of operational efficiency in healthcare and why is it important for providers to pay attention to it. We then looked into two important areas of automation in healthcare – clinical document processing and voice-based applications. Each of these areas consumes a lot of time, as they require manual intervention for processing and an understanding of the clinical information embedded within them. We looked at some common methods of automating the extraction of clinical information from these unstructured data modalities and processing them to create a longitudinal view of a patient, a vital asset to have for the applications of clinical analytics and ML.

Lastly, we built an example application to transcribe a clinical dictation using Amazon Transcribe Medical and then process that transcription using Amazon Comprehend Medical to extract clinical entities into a structured row or column format.

In *Chapter 5*, *Implementing Machine Learning for Healthcare Payors*, we will look into some areas of applications of ML in the health insurance industry.

Further reading

- **Transcribe Medical** documentation: `https://docs.aws.amazon.com/transcribe/latest/dg/transcribe-medical.html`

- **Comprehend Medical** documentation: `https://docs.aws.amazon.com/comprehend-medical/latest/dev/comprehendmedical-welcome.html`

- **Textract** documentation: `https://docs.aws.amazon.com/textract/latest/dg/what-is.html`

5
Implementing Machine Learning for Healthcare Payors

Health insurance is an integral part of a person's well-being and financial security. Unlike countries such as the UK and Canada, the US does not have a concept of universal healthcare. The majority of US residents are covered by plans from their employers who contract with private health insurance companies. Others rely on public insurance such as **Medicare** and **Medicaid**. The rising cost of healthcare has made it almost impossible for anyone to survive without health insurance. No wonder that, by the end of 2020, over 297 million people in the US had coverage for health insurance, with the number trending higher every year.

Health insurance companies, also known as healthcare **payors**, are organizations that cover the healthcare costs incurred by subscribers to their plan, known as **payees**. The payees (patients) or providers submit a **claim** to the payor for the costs incurred, and the payor, after doing their due diligence, pays out the amount to the beneficiary. For a fixed amount paid at regular intervals, known as a **premium**, the subscribers get access to coverage of healthcare costs such as preventive care visits, labs, medical procedures, and prescription drugs. The costs that they pay out of pocket and the healthcare costs that are covered depend on the health insurance plan, or simply the **health plan**. The health plan is a package of charges and services that the payor provides as a choice to their subscribers. Most of these health plans come with a **deductible** amount that needs to be paid out of pocket before the cover kicks in; the amount of the deductible is decided by the plan you pick.

In this chapter, we will look into the details of how a health insurance company processes a claim, which is the largest **Operational Expense (OpEx)** for a payor. We will look at the different stages of claim processing and the areas of optimizing the claim processing workflow. These optimizations are driven by ML models that can automate manual steps and find hidden patterns in claims data, from which operational decisions can be made. Then, we will become familiar with SageMaker Studio, the ML **integrated development environment** (IDE) from AWS, and use it to build an example model for predicting the claim amount for Medicare patients. In this chapter, we will cover the following topics:

- Introducing healthcare claims processing
- Implementing machine learning in healthcare claims processing workflows
- Introducing Amazon SageMaker Studio
- Building an ML model to predict Medicare claim amounts

Technical requirements

The following are the technical requirements that you need to complete before building the example implementation at the end of this chapter:

1. Complete the steps to set up the prerequisites for Amazon SageMaker, as described here: `https://docs.aws.amazon.com/sagemaker/latest/dg/gs-set-up.html`.

2. Create an **S3** bucket, as described in *Chapter 4*, section *Building a smart medical transcription application on AWS* section, under the *Creating an S3 bucket*. If you already have an S3 bucket, you can use that instead of creating a new bucket.

3. Onboard to SageMaker Studio Domain using the quick setup, as described here: `https://docs.aws.amazon.com/sagemaker/latest/dg/onboard-quick-start.html`.

> **Note**
>
> If you have already onboarded a SageMaker Studio domain from a previous exercise, you do not need to perform *step 3* again.

4. Once you are in the SageMaker Studio interface, click on **File | New | Terminal**.

5. Once in the terminal, type the following command:

```
git clone https://github.com/PacktPublishing/Applied-
Machine-Learning-for-Healthcare-and-Life-Sciences-using-
AWS.git
```

You should now see a folder named `Applied-Machine-Learning-for-Healthcare-and-Life-Sciences-using-AWS`.

> **Note**
>
> If you have already cloned the repository in a previous exercise, you should already have this folder. You do not need to do *step 5* again.

6. Familiarize yourself with the SageMaker Studio UI components: `https://docs.aws.amazon.com/sagemaker/latest/dg/studio-ui.html`.

Once you have completed the preceding steps, you should be all set to execute the steps in the example implementation in the final section of this chapter.

Introducing healthcare claims processing

Claims are at the heart of how money is exchanged between healthcare payors and the provider's ecosystem. A claim decides the amount that a provider gets for the healthcare services they perform. There are multiple stages in the life cycle of a claim from the time it is submitted, adjudicated, and disbursed. This process is known as claims processing, and the associated workflow is known as a claims processing workflow.

Before we look into the details of claims processing stages, it is important to understand what a claim typically contains. The claim has some basic information about the patient in question. It has the details of the healthcare services that the patient received and the charges for those services as quoted by the provider. If the payor agrees with the charges, it pays the amount to the provider. Most of these claims are paid out by the payor directly to the provider, but there are some instances when a patient needs to be involved. The most common reason for this is the health plan that the patient is enrolled in doesn't fully cover the cost of the healthcare service they received, in which case the provider sends an invoice for the remaining amount to the patient to be paid directly to the provider. This completes the full life cycle of the claim. The claims processing workflow consists of multiple steps that might vary from payor to payor. However, the basic workflow stages consist of some key steps that are consistent. Now, let us look at these steps of claims processing, as shown in the following diagram:

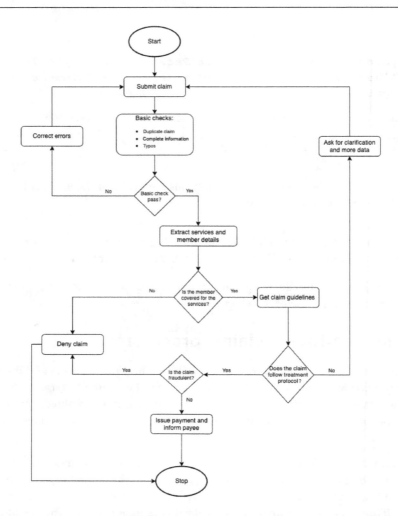

Figure 5.1 – Claims processing workflow

Now, let us describe the steps in more detail:

1. The provider submits an electronic or a paper claim to the health insurance company. Sometimes, it might be a claim submitted directly by the patients for the services they have already paid for to reimburse themselves.

2. The first step that the payors do is to check the claim for completeness and accuracy. For example, are all the mandatory fields complete, are there any typos in the claim, is the data accurate, and is the claim within the prescribed time deadline? If not, the claim is sent back for corrections and resubmission.

3.   The next step is to extract services and member details. The member details are used to verify the status of the member with the health insurance company and also if they are covered for the services being claimed per the health plan they are enrolled in. If not, the claim is denied and sent back with reasons for denial.

4.   The next step is to find out if the claim is per the guidelines. These are broad-ranging guidelines to check things such as whether the services are needed in the context of the patient's history, are the services safe for the patient, and whether the service will improve the patient's condition. This is to ensure that the patients are not billed for the care they don't need. If not, the adjudicator sends the claim back to get more data and clarification about the claim or the associated amounts.

5.   The next step is to find out if there is a risk of fraud in the claim using a set of predefined criteria. If the claim is determined to be fraudulent, it is denied.

6.   Finally, if all checks pass, the claim amount is paid and the subscriber is sent a statement, known as the **explanation of benefits** for their records.

Now that we have a good understanding of the claims processing workflow, let us look at how we can optimize it by applying ML at various stages of the workflow.

Implementing ML in healthcare claims processing workflows

Health insurance companies make money by collecting premiums from their subscribers. Hence, the more subscribers they have, the better it is for their business. However, the larger the subscriber base, the greater the number of claims to process. Not to mention the complexities as a result of managing large health plans and choices. These introduce bottlenecks in claim processing. The good news is that most of the steps in claims processing, as highlighted in the previous section, are repetitive and manual. This provides us with the opportunity to automate some of these tasks using ML. Now, let us look at a few examples of the application of ML in claims processing:

- **Extracting information from claims**: There is still a large volume of claims that get submitted as paper claims. These claims require humans to parse information from the claim forms and manually enter the details of the claims in a digital claim transaction system to be stored in a database. Moreover, the extracted information is also manually verified for accuracy and completeness. By utilizing **optical character recognition** (**OCR**) software, you can automate the extraction of information from claim forms. OCR technology is driven by **computer vision models** trained to extract information from specific fields, sections, pages, or tables in a claim form. The information extracted from these models is easily verifiable for completeness and accuracy using simple rules that can be programmatically run and managed. The AWS service **Textract** is an OCR service that allows you to extract information into PDFs and images and provides APIs that enable you to integrate Textract into a claims processing workflow.

- **Checking whether the claim is per guidelines**: Checking if the claim is compliant and is per guidelines requires a deeper analysis of the claim data. Usually, this is not possible using a rules-based approach and requires the application of **natural language processing (NLP)** algorithms on claims data. For example, one of the key steps in checking a claim's compliance is to extract clinically relevant entities from it. A **named entity recognition (NER)** model can extract key entities from the claim data such as diagnoses, medications, test names, procedure names, and more, removing the need to manually extract such information. **Amazon Comprehend Medical** is a service that does just that. Sometimes, these entities might need to be combined with other features such as the demographic details of the patient, the provider details, and the historical information about the patient's conditions and past claims. This combined view can provide new insights into the claim, and ML algorithms can learn from these features to check compliance with specific guidelines.

- **Checking the risk of fraud**: Fraud detection in healthcare claims is one of the most common use cases. It involves ingesting claim records from multiple data sources and using them to identify anomalies in these transactions. You can approach this as a supervised or unsupervised learning problem. In the supervised setting, the algorithm can learn from labeled data to classify a transaction as fraudulent. In an unsupervised setting, the algorithms can point out how much a particular transaction differs from the others and flag them as anomalous. Both these approaches are designed to remove manual intervention and lower the rate of claim denials due to fraud, which is a huge overhead for healthcare payors.

- **Financial trends and budgeting**: With the cost of healthcare increasing every year, healthcare payors need to regularly evaluate their plan premiums and claim amounts to make sure they are profitable. ML can identify trends and patterns in the financial data in claims and predict future forecasts and costs. This allows payors to revise plan structures and premium costs for future subscribers.

Now let us use ML to build a model to predict claim costs. Before we do that, we will dive into SageMaker Studio, a service from AWS that we will use to build the model.

Introducing Amazon SageMaker Studio

Amazon SageMaker Studio is an IDE for ML that is completely web-based and fully managed by AWS resources in the background. It provides an intuitive interface to access ML tools to build, train, deploy, monitor, and debug your ML models. It also provides studio notebooks, which have a **JupyterLab** interface preinstalled with popular data science libraries that allow you to begin experimenting immediately upon getting access to the studio notebooks interface. These notebooks can be scaled up or down for CPUs or GPUs depending on the workloads you want to run on them and also provides terminal access for you to install and manage third-party libraries for local runs of your experiments. Once you are done experimenting locally, you can call multiple SageMaker jobs to scale out your experiments to larger datasets and workloads that can be horizontally scaled to multiple instances instead of a single notebook environment. For example, **SageMaker Processing** jobs can scale out and preprocess training data for feeding into a **SageMaker Training** job that can then train a model on large-scale datasets. The trained model can then be deployed as a **SageMaker endpoint** for real-time inference via an API call. This workflow can be templatized and maintained as a pipeline using **SageMaker Pipelines** to make it repeatable and managed as a production **MLOps pipeline**.

SageMaker tools

There are multiple ML tools available for access from within the Sagemaker Studio interface that makes integrating them into your workflows easy. You can refer to the complete list of features related to Sagemaker Studio in the SageMaker Studio developer guide here: `https://docs.aws.amazon.com/sagemaker/latest/dg/studio.html`

In this chapter, we will touch upon two aspects of the SageMaker Studio IDE, which we will use in the example implementation in the next section, SageMaker Data Wrangler and SageMaker Studio notebooks.

SageMaker Data Wrangler

SageMaker Data Wrangler allows you to create preprocessing data workflows for your downstream ML training jobs. ML algorithms need data to exist in a certain format before you can start training. SageMaker Data Wrangler provides a visual workflow building interface known as a data flow to design and manage different transformations to your data as you make it suitable for ML algorithms. You can then choose to analyze your transformed data or export the data into S3. Additionally, you have options to templatize the processing using a processing job or a SageMaker pipeline. Let us look at some of these features in more detail.

Importing your data

You can import data into data wrangler from a variety of AWS sources such as S3, Athena, and Redshift. You can also import data into Data Wrangler from non-AWS sources such as Data Bricks (using a JDBC driver) and Snowflake. Depending on the choice of your data source, you can provide the required parameters for Data Wrangler to authenticate and pull data from those sources.

Once imported, the dataset automatically appears in the data flow screen as the starting point of a data flow. Data Wrangler samples the data in each column of your dataset and infers the data types for them. You can change these default datatypes by going to the **data types** step in your data flow and choosing **Edit data types**. Additionally, Data Wrangler maintains a backup of the data imported from Athena and Redshift in the default S3 bucket that SageMaker creates for you. This bucket name is formatted as a *sagemaker-region-account number*. Once imported, you are now ready to transform and analyze your data.

Transforming and analyzing your data

The Data Wrangler data flow screen allows you to join multiple datasets together. You can generate insights from the joined dataset by generating a data quality and insights report. The report provides important information such as missing values and outliers in your dataset, which, in turn, allows you to choose the right transformations to apply to your data. Additionally, you can choose a target column in your dataset and create a quick model to determine whether you have a good baseline upon which you would want to improve. It also gives you feature importance and column statistics for the features. These are important insights to have at the beginning of the ML life cycle process to have an understanding of how good the dataset is for the problem at hand. Once you are done with the analysis, you can proceed to transform your data using a variety of built-in transformations. These range from data cleansing transforms such as dropping columns and handling columns with missing values to more advanced transforms for encoding text and categorical variables. Data Wrangler also allows you to write your own transformation logic using Python, PySpark, or Pandas. These are great for complex transforms that cannot be handled by built-in transformations. Each addition of a transform updates the data flow diagram with new steps so that you can visualize your transformation pipeline in real time as you build it. Once done transforming, Data Wrangler allows you to visualize your data by generating data visualization analysis such as histograms and scatter plots. You can extend these visualizations with your own custom code, which provides even more options for generating rich visual analysis from the Data Wrangler analysis interface.

Exporting your data and workflows

Once you have completed your data transformation and analysis steps, you are ready to begin training an ML model with this dataset. Data Wrangler provides multiple options for you to export your data transformations. You can export the data into an S3 location. Also, you can choose to export to SageMaker pipelines to templatize the data processing flow. This option is ideal for large-scale production pipelines that need to standardize the process of data preprocessing. Another option provided by Data Wrangler is the ability to export the data into SageMaker Feature Store. SageMaker Feature Store is a centralized feature repository that you can share and reuse for all your downstream training needs. By exporting the transformed data into SageMaker Feature Store, you make the data discoverable so that an audit trail can be maintained between the training job and the raw feature set. Lastly, you can export the transformations as Python code. This Python is ready to use and is autogenerated by Data Wrangler. You can integrate this code into any data processing workflow and manage the pipeline yourself.

Data Wrangler

To learn more about SageMaker Data Wrangler, you can refer to the developer guide here: `https://docs.aws.amazon.com/sagemaker/latest/dg/data-wrangler.html`

SageMaker Studio notebooks

ML is an iterative process, and a preferred way to iterate through ML problems is in a notebook environment. **SageMaker Studio notebooks** provide data scientists with a fully managed JupyterLab notebook interface, which is a popular notebook environment. The notebook environment of choice is configurable. You have the option of selecting from a set of EC2 instance types that basically define the hardware (CPU versus GPU and memory) your notebook will run on. Some of these instances are designated as fast launch instances, which can be launched in under two minutes. The notebooks are backed by persistent storage that lies outside the life of the notebook instances, so you can share and view notebooks even if the instance it runs on is shut down.

Additionally, you can define a SageMaker image, which is a container image that's compatible to run in SageMaker Studio. The image defines the kernel and packages that will be available for use in the notebook environment. There are multiple pre-built images with packages to choose from, including `scikit-learn`, `pandas`, `MxNet`, `NumPy`, and `TensorFlow`. You also have the option of bringing your own image into SageMaker Studio, which allows you flexibility and granular customization should you need it.

SageMaker Studio notebooks allow you to share notebooks with others using a snapshot. The snapshot captures the environment and its dependencies so that the user you share it with will be able to reproduce your results easily. Your notebooks come with Terminal access, which gives you the flexibility of installing third-party packages using package managers such as `Conda` and `pip`.

Now that we have a good idea about SageMaker Studio features and have an overview of Data Wrangler and Studio notebooks, let us put our knowledge to use. In the next section, we will build an ML model to predict the Medicare payment costs for a group of subscribers.

Building an ML model to predict Medicare claim amounts

Medicare is a benefit that is provided for people in the US who are aged over 65 years. In general, it is designed to cover healthcare costs for seniors including procedures, visits, and tests. The cost of Medicare is covered by the federal government in the US. Just like with any private insurance, they need to analyze the data to find out ways to estimate payment costs and make sure they are setting aside the right budget when compared to the premiums and deductible amounts. This is important as the cost of healthcare services changes over time. ML can learn from past claim amounts and predict claim amounts for new subscribers of the plan. This can help the insurance provider to plan for future expenses and identify areas for optimization. We will now build this model using Amazon SageMaker. The goal of this exercise is to create an end-to-end flow for feature engineering using SageMaker Data Wrangler and a model in the SageMaker Studio notebook to predict the average Medicare claim amount.

Acquiring the data

The dataset used for this exercise is called **Basic Stand Alone (BSA) Inpatient Public Use Files (PUF)**. It is available from the CMS website at `https://www.cms.gov/Research-Statistics-Data-and-Systems/Downloadable-Public-Use-Files/BSAPUFS/Inpatient_Claims`.

The dataset consists of a PUF for Medicare inpatient claims from the year 2008. It has a 5% random sample of such beneficiaries who are randomly selected. The dataset consists of demographic information such as age and gender, the **diagnosis-related group (DRG)** for the claim, the ICD-9 code for the primary procedure, the length of stay for the inpatient visit, and an average claim amount. The claim amount is divided into approximate quantiles for a base DRG group and the average amount of that quantile is captured. For the purposes of this exercise, we will use this average amount as our prediction target, making it a regression problem. You can go over the details of the dataset and associated PDF files that explain the data variables and schema in more detail at the preceding link:

1. Once you have a good understanding of the data, go ahead and download the file 2008 BSA Inpatient Claims PUF (ZIP) file from the following link: `https://www.cms.gov/Research-Statistics-Data-and-Systems/Downloadable-Public-Use-Files/BSAPUFS/Downloads/2008_BSA_Inpatient_Claims_PUF.zip`

2. Unzip the downloaded file, which should result in a CSV file named `2008_BSA_Inpatient_Claims_PUF.csv`.

3. We will also download the dictionary files to help us understand the codes used in the data for attributes like DRG and Procedures. You can download the dictionary zip file from the following location: `https://github.com/PacktPublishing/Applied-Machine-Learning-for-Healthcare-and-Life-Sciences-using-AWS/blob/main/chapter-05/dictionary.zip`

4. Unzip the file, which will create a folder named `Dictionary` with the dictionary CSV files in it.

5. Navigate to the S3 bucket in the AWS management console and upload the files you just downloaded. You can follow the steps described here if you have any issues with this: `https://docs.aws.amazon.com/AmazonS3/latest/userguide/upload-objects.html`

Now you are ready to proceed with SageMaker to begin the feature engineering step using SageMaker Data Wrangler.

Feature engineering

As discussed earlier, we will be using SageMaker Data Wrangler for the feature engineering step of this exercise. The raw data and the dictionary files we uploaded on S3 are the sources for this step, so make sure you have them available:

1. Navigate to the SageMaker service on the AWS management console, and click on **Studio** in the left-hand panel:

Figure 5.2 – The AWS console home page for Amazon SageMaker

2. On the next screen, click on the **Launch SageMaker Studio** button. This will open the user list. Click on the **Launch app** drop-down menu and then **Studio**. This will open the Studio interface, which will have a default launcher, as shown in the following screenshot:

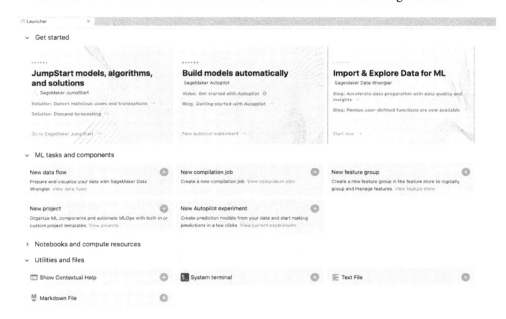

Figure 5.3 – The SageMaker Studio launcher

> **Note**
>
> If you have any trouble with the opening of the studio, please go back and review the prerequisite steps and then come back to this step.

3. In the **Explore and prepare data for ML** tile, click on **Start now**. This will open a page that will show a **Data Wrangler is loading** message. Wait for this step to complete, as it will take a few minutes. This delay is only for the first time you launch Data Wrangler.

4. Once Data Wrangler is ready, you should see a `.flow` file called `untitled.flow`. You can change the name of this file to a more relevant name such as `claimsprocessing.flow` in the file explorer on the left-hand panel of SageMaker studio by right-clicking on the file name.

5. In the import tab of `claimsprocessing.flow`, select Amazon S3. Navigate to the `2008_BSA_Inpatient_Claims_PUF.csv` file. This will open up a preview panel with a sample of your claims data. Click on the **Import** button to import this file:

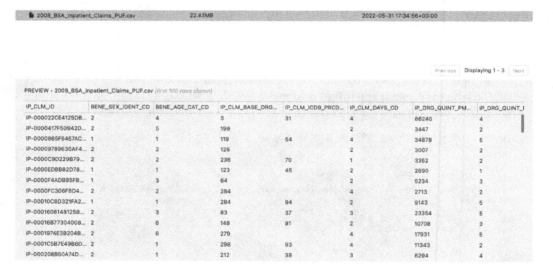

Figure 5.4 – Data preview screen in SageMaker Data Wrangler

6. On the next screen with the data flow diagram, double-click on the **Data types** step, which will open up the data types inferred by Data Wrangler for the columns. Except for the first column, **IP_CLM_ID**, which is alphanumeric, the rest of the columns are all integers. You can change them to **Long**. Select **Long** from the drop-down menu, as shown in the following screenshot, and click on **Preview** and then **Update**:

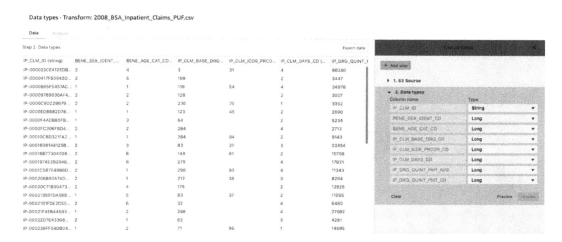

Figure 5.5 – The data type step in the Data Wrangler flow

7. Click on **Add step** and scroll down to the transform option, **Manage columns**. Make sure **Drop column transform** is selected in the transform drop-down list. Select **IP_CLM_ID** from the **columns to drop** drop-down list. This column is a unique identifier having no significance to ML, so we are going to drop it. Click on **Preview** and then **Add**:

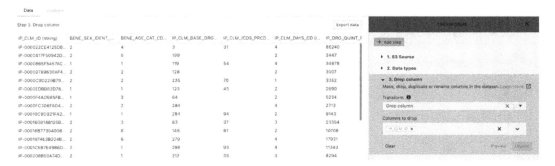

Figure 5.6 – Dropping the column transform in SageMaker Data Wrangler

8. Next, we will import two more files into Data Wrangler from our dictionary. We will join these files with 2008_BSA_Inpatient_Claims_PUF.csv to bring some text attributes into our data. Click on **Back to data flow** and follow *step 5* two more times to import the DiagnosisRelatedGroupNames.csv and InternationalClassification_ OfDiseasesNames.csv files into Data Wrangler.

9. Next, repeat *step 6* to change the data type of the integer column in both datasets to **Long**. You are now ready to join the datasets.

10. Click on the + sign next to the drop column step in the `2008_BSA_Inpatient_Claims_PUF.csv` data flow. Then, click on **Join**:

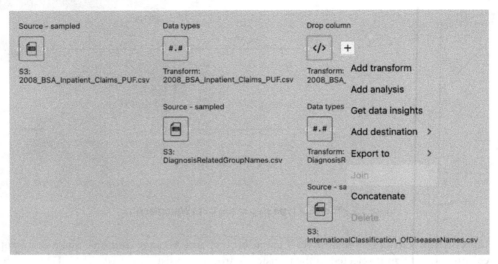

Figure 5.7 – The Join option in SageMaker Data Wrangler

11. On the **Join** screen, you will see a panel where you can see the left-hand dataset already selected. Select the right-hand dataset by clicking on the **Data types** step next to `DiagnosisRelatedGroupNames.csv`. Next, click on the **Preview** tab:

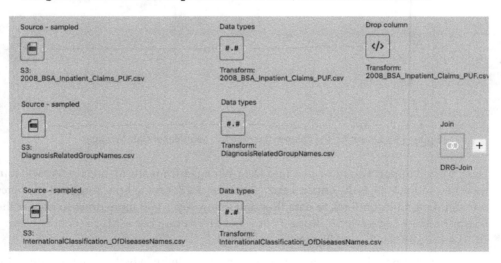

Figure 5.8 – The Join data flow step in SageMaker Data Wrangler

12. On the **Join** preview screen, type in a name for your join step such as DRG-Join, and select the join type as **Inner**. Select the **IP_CLM_BASE_DRG_CD** column for both the left and right columns to join. Then, click on **Preview** and **Add.**

13. Now we will join the InternationalClassification_OfDiseasesNames.csv data. Click on the + button next to the **Join** step in your data flow and repeat *steps 10, 11,* and *12.* Note that this time, the columns to join on the left and right will be **IP_CLM_ICD9_PRCDR_CD.**

14. As a result of the joins, there are some duplicate columns that are added to the dataset. Also, since we have long descriptive columns for DRG and procedures, we can remove the original code columns that are integers. Click on the + sign after the second join step and click on **Add transform**. Follow step to drop the **IP_CLM_ICD9_PRCDR_CD_0, IP_CLM_ICD9_ PRCDR_CD_1, IP_CLM_BASE_DRG_CD_0,** and **IP_CLM_BASE_DRG_CD_1** columns.

15. We are now done with feature engineering. Let us export the data to S3. Double-click on the last **Drop column** step and click on **Export data**. On the next **Export data** screen, select the S3 bucket where you want to export the transformed data file in CSV format. Leave all the other selections as default and click on **Export data**. This will export the CSV data file to S3. You can navigate to the S3 location and verify that the file has been exported by clicking on the S3 location hyperlink:

Figure 5.9 – The Export data screen in SageMaker Data Wrangler

Now that we are ready with our transformed dataset, we can begin the model training and evaluation step using SageMaker Studio notebooks.

Building, training, and evaluating the model

We will use a SageMaker Studio notebook to create a regression model on our transformed data. Then, we will evaluate this model to check whether it performs well on a test dataset. The steps for this are documented in the notebook file, which you can download from GitHub at `https://github.com/PacktPublishing/Applied-Machine-Learning-for-Healthcare-and-Life-Sciences-using-AWS/blob/main/chapter-05/claims_prediction.ipynb`:

1. Open the Sagemaker Studio interface.

2. Navigate to the path `Applied-Machine-Learning-for-Healthcare-and-Life-Sciences-using-AWS/chapter-5/`. and open the file `claims_prediction.ipynb`.

3. Read the notebook and complete the steps, as documented in the notebook instructions.

4. By the end of the notebook, you will have a scatter plot showing a comparison between actual and predicted values, as shown in the following screenshot:

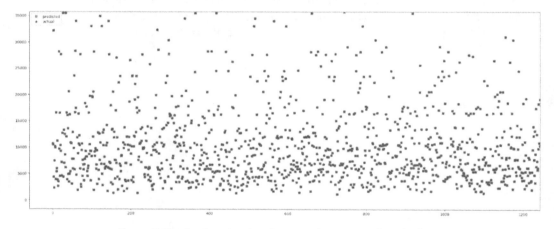

Figure 5.10 – Scatter plot showing actual versus predicted values

This concludes our exercise. It is important to shut down the instance on which Data Wrangler and the Studio notebooks are running to avoid incurring charges. You can do so by following the instructions at the following links:

- Shutting down Data Wrangler: `https://docs.aws.amazon.com/sagemaker/latest/dg/data-wrangler-shut-down.html`

- Shutting down Studio: `https://docs.aws.amazon.com/sagemaker/latest/dg/studio-tasks-update.html`

Summary

In this chapter, we dove into the details of the healthcare insurance industry and its business model. We looked at claim processing in detail and went over different stages of how a claim gets processed and disbursed. Additionally, we looked at the various applications of ML in the claims processing pipeline and discussed approaches and techniques to operationalize large-scale claims processing in healthcare. We got an introduction to SageMaker Studio and, more specifically, SageMaker Data Wrangler and the SageMaker Studio notebooks. Lastly, we used our learnings to build an example model to predict the average healthcare claims costs for Medicare patients.

In *Chapter 6, Implementing Machine Learning for Medical Devices and Radiology Images*, we will dive into how medical images are generated and stored and also some common applications of ML techniques in medical images to help radiologists become more efficient. We will also get introduced to the medical device industry and how the ML models running on these devices can help with timely medical interventions.

6

Implementing Machine Learning for Medical Devices and Radiology Images

Visualizing the human body using imaging studies is a vital part of a patient's care journey. The generation, storage, and interpretation of these images are widely known as **medical imaging**. There are multiple types of medical images that address different diagnostic needs of patients. Some of the common types include X-ray, ultrasound, **computerized tomography** (**CT**), and **magnetic resonance imaging** (**MRI**). Each of these types of imaging studies needs specific equipment, lab setup, and trained professionals who can operate the equipment. Radiologists are medical professionals who specialize in diagnosing medical conditions using medical images. A radiologist also performs imaging studies using specialized equipment and is able to generate and interpret radiology reports. The rate of growth in imaging studies has continued to rise since the early 2000s. However, since these studies are costly, it is important to exercise caution when prescribing them and they should only be performed when really necessary. Moreover, ordering unnecessary imaging studies will add to individuals' workloads, and may cause radiologists to get burned out.

ML can help radiologists become more efficient in the interpretation of medical images. Computer vision models can interpret the information in medical images and support radiologists by helping them make clinical decisions faster and more accurately. In addition, radiology equipment such as MRI machines and CT scanners are also getting smarter through the use of ML. Models embedded directly onto these devices allow them to perform better and triage image quality locally even before a radiologist looks at it. This pattern is common in a large category of medical devices, such as glucometers and **electrocardiogram** (**ECG**), that utilize the **Internet of Things** (**IoT**) to run ML-based inference either locally or on data aggregated from these devices in a central data store.

In this chapter, we will get an understanding of medical devices and how smart connected devices are revolutionizing healthcare. We will then dive into the details of the medical imaging system components. We will look at the different components in the workflow and understand the role that each of those components plays in the overall workflow. We will also look at how ML can optimize medical device workflows. Finally, we will use Amazon SageMaker to build a medical image classification model. These topics are shown in the following sections:

- Introducing medical devices
- Introducing radiology imaging system components
- Applying ML to medical devices and radiology imaging
- Introducing SageMaker training
- Building a medical image classification model using SageMaker

Technical requirements

The following are the technical requirements that you need to complete before building the example implementation at the end of this chapter:

1. Complete the steps to set up the prerequisites for Amazon SageMaker as described here: `https://docs.aws.amazon.com/sagemaker/latest/dg/gs-set-up.html`.

2. Create a SageMaker notebook instance by following the steps in the following guide: `https://docs.aws.amazon.com/sagemaker/latest/dg/howitworks-create-ws.html`.

> **Note**
> At the step where you need to choose the notebook instance type, please select **ml.p2.xlarge**.

3. Create an S3 bucket, as described in *Chapter 4*, in the *Building a smart medical transcription application on AWS* section, under *Create an S3 bucket*. If you already have an S3 bucket, you can use that instead of creating a new bucket.

4. Open the Jupyter notebook interface of the Sagemaker notebook instance by clicking on Open Jupyter link on the Notebook Instances screen.

5. On the top right corner of the Jupyter notebook, click on **New** and then **Terminal**.

6. Type the following commands on the terminal screen:

```
$cd Sagemaker
$git clone https://github.com/PacktPublishing/Applied-
Machine-Learning-for-Healthcare-and-Life-Sciences-using-
AWS.git
```

You should now see a folder named `Applied-Machine-Learning-for-Healthcare-and-Life-Sciences-using-AWS`.

> **Note**
>
> If you have already cloned the repository in a previous exercise, you should already have this folder. You do not need to do *step 6* again.

Once you have completed these steps, you should be all set to execute the steps in the example implementation in the last section of this chapter.

Introducing medical devices

The **medical device** industry comprises a wide range of organizations that do research, development manufacturing, and distribute devices and associated technologies that help prevent, diagnose, and treat medical conditions. It's important to note that the term medical device has wide applicability. It includes things you normally get over the counter, such as a Band-Aid or a testing kit, but it also includes highly specialized instruments, such as CT scanners and respirators. The key differentiator for a medical device is that it achieves its purpose due to its physical structure instead of a chemical reaction.

The US medical devices industry accounts for billions of dollars in exports, which has continued to rise over the years. In the US, the **Food and Drug Administration (FDA) Center for Medical Devices and Radiological Health (CDRH)** is the federal regulating agency that authorizes the use of medical devices. It also monitors adverse events associated with medical device use and alerts consumers and medical professionals as needed. The process of regulating a medical device depends on its classification. Let's look at the different classes of medical devices as defined by the FDA.

Classes of medical devices

Due to the wide applicability of the term, medical devices are categorized into different classes based on the risk they pose. There are three classes for a medical device:

- **Class 1 medical device**: This is the lowest-risk medical device and is generally safe for consumer use, provided a set of safety protocols and guidelines are followed. Examples of class 1 devices include surgical masks, bandages, and latex gloves.

- **Class 2 medical device**: A class 2 medical device is known to have an intermediate risk and users need to follow special guidelines and controls. These include premarket review and clearance by the FDA after an elaborate review of the guidelines. Examples of class 2 medical devices include pregnancy test kits, blood transfusion kits, and syringes.

- **Class 3 medical device**: Class 3 medical devices are high-risk devices that support life or may be implanted into one's body. This class is highly regulated by the FDA and must follow the FDA's class 3 regulations, which include the most rigorous review process. Examples of class 3 medical devices include pacemakers, defibrillators, and different types of implants (such as breast implants or prosthetic implants).

All medical devices must be registered and listed with the FDA. To make it easy for consumers to find the appropriate classification of a medical device, the FDA provides a classification database (`https://www.accessdata.fda.gov/scripts/cdrh/cfdocs/cfpcd/classification.cfm`) to search using the part number or the name of the device. To learn more about the classification of a medical device, visit the following link: `https://www.fda.gov/medical-devices/overview-device-regulation/classify-your-medical-device`.

In the US, the CDRH determines the approval process for the medical device, depending on the classification it belongs to. The process may include steps such as the submission of premarket approval and clinical evidence via a clinical trial that has evidence that the device is safe to be used on humans.

Let us now look at a special category of medical devices that involve computer software, also known as **software as a medical device (SaMD)**.

Understanding SaMD

Software forms an integral part of a modern medical device. The software could be a part of the medical device. It may also be related to the support, maintenance, or manufacturing of the medical device itself. A third category of software related to medical devices is known as **SaMD**. SaMD is a type of software intended to be used for medical purposes without being part of the hardware of the medical device. It provides a wide-range of applications for medical technology, from off-the-self solutions to custom-built solutions for specific scenarios.

The FDA has been working over the last few years to develop or modernize regulatory standards and the approval process to cater to the needs of the digital health market enabled via the use of SaMD. In 2017, the FDA released new guidelines for evaluating SaMD applications that propose an accelerated review of digital health products. The revised framework for the approval of SaMD proposes a **regulatory development kit (RDK)** containing tools to clarify the requirements at each stage of the approval process and provide templates and guidance for frequently asked questions. It also proposes a precertification program to accelerate the approval of SaMD for organizations that meet certain criteria of excellence. To learn more about the clinical evaluation of SaMDs, visit the following link: `https://www.fda.gov/regulatory-information/search-fda-guidance-documents/software-medical-device-samd-clinical-evaluation`.

In this section, we were introduced to medical devices and how they are regulated for use. One of the common uses of medical devices is in the radiology department. This department is responsible for generating, storing, and interpreting different modalities of radiology images. Let's understand the different components of a radiology imaging system and the role they play in the imaging workflow.

Introducing radiology imaging system components

The need for medical imaging in a patient's care journey depends on the specialty that is treating them. For example, a cardiology department may need a medical image-driven diagnosis at a different point in time in patient care from the gastroenterology department. This makes standardizing the medical imaging workflow really difficult. However, there are some standard components that are common across all medical imaging systems. These components work in sync with each other to create a workflow specific to the specialty in question. Let us look at these components in more detail:

- **Image acquisition**: Medical images are generated from specialized equipment located in clinical facilities. The equipment and the facility where it is located depend on the type of medical image it generates. The format and size of images are also dependent on the type. The most commonly used standard for storing and sharing medical images is **Digital Imaging and Communications in Medicine (DICOM)**. DICOM provides an interoperability standard for medical imaging equipment so the different pieces of equipment can communicate with each other. The DICOM standard provides a file format and communications protocol for exchanging medical images. To learn more about the DICOM format, check out the following link: `https://dicom.nema.org/medical/dicom/current/output/html/part01.html`.

- **Image data storage and management**: Once the medical images are acquired from this equipment, they need to be centrally managed and stored in a storage system. This storage and management system for medical images is known as a **picture archiving and communications system (PACS)**. PACS is at the heart of a medical imaging workflow. Not only does it provide a central location for the storage of images and metadata, but it also provides capabilities to authenticate users and manage access to the medical images. It integrates with downstream systems to provide analysis and visualization capabilities. You can search through the medical image repository in a PACS using DICOM metadata and tags to retrieve historical imaging records for a particular patient.

- **Medical image visualization and analysis**: To interpret the details in medical images, radiologists need to visualize them using medical image visualization tools. There are multiple visualization tools available on the market. Some of them come integrated with PACS, so radiologists can go from storing and interpreting seamlessly. The visualization tools come inbuilt with capabilities to render two-dimensional and three-dimensional medical image modalities stored in standard formats such as DICOM. They also provide radiologists with tools to annotate the medical images by highlighting a certain area of the image or attaching a label to the image for downstream analysis. For example, these analysis steps could include the creation of aggregated views for a patient's imaging studies over a period of time or searching through the image repository for images with certain clinical conditions.

Now that we understand the key components of the medical imaging system and the various associated medical devices, let us dive into how ML can be applied to such workflows.

Applying ML to medical devices and radiology imaging

Unlike traditional software, ML models evolve over time and improve as they interact with real-world data. Also, due to the probabilistic nature of the models, it is likely that the output of these models will change as the statistics behind the data shift. This poses a challenge in applying these models for regulated medical workflows because the medical decision-making process needs to be consistent and supported by the same evidence over and over again. Moreover, the results of an ML model aiding in a clinical decision-making process need to be explainable. In other words, we cannot treat the model as a "black box"; we need to understand its inner workings and explain its behavior in specific scenarios.

In spite of these challenges, the FDA recognizes that AI/ML has the potential to transform healthcare due to its ability to derive insights from vast amounts of data generated in healthcare practice every day. In 2019, the FDA published a discussion paper (https://www.fda.gov/media/122535/download) that proposes a regulatory framework for modifications to AI/ML-based SaMD. This is a welcome initiative as it recognizes the need for modernization in medical devices and associated software. Let us now look at a few examples of how ML can be applied to medical devices and radiology imaging:

- **Detecting failures and operational continuity**: ML models can work on recognizing trends in the operational metrics or telemetry data gathered from medical devices. These data points can have hidden indicators of device failures or malfunctions that can cause disruption in the devices' operations. Using data gathered from sensors mounted on the device that continuously stream data to ML models using IoT gateways, the models can recognize trends in the data and alert users about upcoming unwanted events so timely corrective actions can be taken to prevent such events. This ensures the operational continuity of the device.

- **Real-time decision-making**: Another application of ML on streaming data from medical devices is to alert patients and medical professionals about trends in a patient's vital signs, such as heart rate, glucose, or blood pressure. The data is gathered at regular intervals from these devices mounted on patients and is used to infer the condition of the patient in real time. The models can also act as a notification system to alert emergency responders in case of emergencies such as falls or heart failure.

- **Aggregated data analysis**: While real-time decision-making is critical, it is also important to analyze aggregated data collected from medical devices to understand trends in clinical data and make business decisions. The data collected from medical devices is typically streamed to central aggregation servers. These servers can run ML models for a variety of tasks, from risk stratification of patients and forecasting long-term adverse events to applying search algorithms to index and tag clinical documents.

- **Radiology image classification, segmentation, and search**: Medical image-based diagnosis can be augmented by computer vision algorithms that are trained to identify clinical conditions from visual biomarkers. Radiology image classification algorithms can categorize radiology images into disease categories or identify anomalous versus normal images. Radiology image segmentation models or object detection models can identify the regions on a medical image that may be of interest to the radiologist. For example, these models can highlight a tumor in a brain CT image to direct the attention of the radiologist. It can also identify specific regions of an image to aid in understanding the anatomy of the patient. Search algorithms can help catalog and identify similar images that demonstrate the same clinical conditions to help compare the patient to past patients and their care plans.

Now that we have an understanding of the medical device industry, its workflows, and the different components of a radiology imaging system, let us now use SageMaker training to train an image classification model for diagnosing pneumonia. Before we do that, we will revise the basics of SageMaker training so we can better understand the implementation steps.

Introducing Amazon SageMaker training

In *Chapter 5*, you learned about SageMaker Studio and how to process data and train a model using SageMaker Data Wrangler and SageMaker Studio notebooks, respectively. Studio notebooks are great for experimentation on smaller datasets and testing your training scripts locally before running them on the full dataset. While SageMaker Studio notebooks provide a choice of GPU-powered accelerated computing, it is sometimes more cost effective to run the training as a job outside the notebook. It is also an architectural best practice to decouple the development environment from the training environment so they can scale independently from each other. This is where SageMaker training comes in. Let us now understand the basics of SageMaker training.

Understanding the SageMaker training architecture

SageMaker provides options to scale your training job in a managed environment decoupled from your development environment (such as SageMaker Studio notebooks). This is done using a container-based architecture and involves orchestrating batch jobs in containers on SageMaker ML compute instances. Let us look at a diagram to understand how training jobs are run on SageMaker:

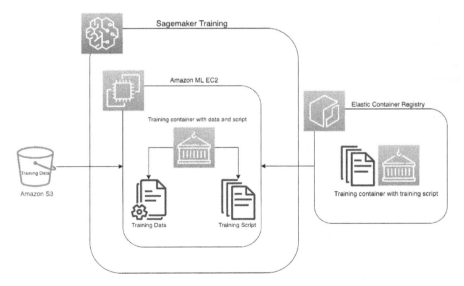

Figure 6.1 – SageMaker training architecture

As shown in *Figure 6.1*, SageMaker training is a container-enabled environment that allows you to run training in a managed training environment. The user provides training data that is stored on S3. In addition, depending on the way you run training (covered in the next section), you may need to provide a Docker URI from the container registry and a training script. The user also provides the SageMaker ML instance type and the count of instances where this training should be run. Once you provide these attributes, SageMaker training downloads the training container image and the code along with the training data on the SageMaker ML instance of choice. It then executes the training following the training script provided. You can monitor the execution of the training job using the status of the training job. During execution, SageMaker outputs the logs from the training job on CloudWatch Logs. Once the model is trained, it is stored on S3 at a predefined path. It can then be downloaded for deployment on your own environment or using SageMaker inference.

Let us now look at various ways in which you can run training on SageMaker.

Understanding training options with SageMaker training

SageMaker provides multiple options when it comes to running training jobs. Here is a summary of the options:

- **Run training using SageMaker-provided algorithms**: SageMaker provides a variety of training algorithms prepackaged into containers. This approach requires you to provide training data to the algorithm in the expected format and the choice of instance you want to run it on. The algorithms are maintained in a Docker container accessible via a Docker URI. SageMaker runs the training using the algorithm and data provided and stores the trained model artifact on S3. For a full list of available SageMaker algorithms, refer to the following link: `https://docs.aws.amazon.com/sagemaker/latest/dg/algos.html`.

- **Run training using a SageMaker framework container with a custom script**: SageMaker provides prebuilt containers for common frameworks such as **TensorFlow**, **MXNet**, **PyTorch**, **scikit-learn**, and **XGBoost**. The framework containers are open source so you can look at what libraries and versions they support. They are also accessible via a Docker URI. You can provide additional libraries or dependencies as a `requirements.txt` file that can be installed during training using `pip`. In addition to the requirements in the previous approach, this approach requires you to provide a training script that SageMaker can execute in your training framework container of choice. You can find the list of framework containers that SageMaker supports at the following link: `https://docs.aws.amazon.com/sagemaker/latest/dg/frameworks.html`.

  Once the training is over, the trained model is stored on S3.

- **Run training on your own container or extend one of the deep learning containers**: This option provides you with the flexibility to run a training job on SageMaker by providing an existing container image or extending one of the prebuilt SageMaker container images.

SageMaker deep learning container images are open source and available at the following link: `https://github.com/aws/deep-learning-containers/blob/master/available_images.md`.

This option is ideal for highly specialized training jobs that may need you to modify the underlying environment or install specialized libraries. You have control over how you build your container and the format of the input and output. At this point, SageMaker is acting as an orchestrator for your job that you can execute on an instance of your choice.

- **Run training using SageMaker marketplace algorithms**: SageMaker provides a marketplace for third-party providers to register their own algorithms and models. The algorithms address a wide range of problem types involving tabular, text, or image data in areas such as computer vision, natural language processing, and forecasting. You can browse and choose your algorithm of choice and subscribe to it. SageMaker then allows you to train on the subscribed model just as you would train using other options. You can find more details about SageMaker marketplace algorithms at the following link: `https://docs.aws.amazon.com/sagemaker/latest/dg/sagemaker-mkt-find-subscribe.html`.

Now that you have a good understanding of SageMaker training, let us use it to train a computer vision model for medical image classification.

Building a medical image classification model using SageMaker

One common application of ML in medical imaging is for classifying images into different categories. These categories can consist of different types of diseases determined by visual biomarkers. It can also recognize anomalies in a broad group of images and flag the ones that need further investigation by a radiologist. Having a classifier automate the task of prescreening the images reduces the burden on radiologists, who can concentrate on more complex and nuanced cases requiring expert intervention. In this exercise, we will train a model on SageMaker to recognize signs of pneumonia in a chest X-ray. Let us begin by acquiring the dataset.

Acquiring the dataset and code

The `Chest X-Ray Images (Pneumonia)` dataset is available on the Kaggle website here: `https://www.kaggle.com/datasets/paultimothymooney/chest-xray-pneumonia?resource=download`.

The dataset consists of 5,863 chest X-ray images in JPEG format organized into three subfolders (`train`, `test`, and `val`). The images are organized into folders named `NORMAL` and `PNEUMONIA`:

1. Download the dataset from the preceding Kaggle download link. The dataset is around 2 GB in size, so make sure you have a reliable internet connection and adequate space on your computer.

2. Unzip the downloaded file. You should see the images organized into three folders: `train`, `test`, and `val`. Organize the data into the following folder structure on your local computer:

```
data/
        test/
                NORMAL/
                PNEUMONIA/
        train/
                NORMAL/
                PNEUMONIA/
        val/
                NORMAL/
                PNEUMONIA/
```

3. Go to the `data` folder and create a ZIP file called `chest_xray_train.zip` by including the `train` and `val` images. If you are using the Linux command line, you can type the following:

```
zip -r chest_xray_train.zip train val
```

4. Upload the `chest_xray_train.zip` file to the S3 bucket you created as part of the prerequisites. Once uploaded, you can delete the ZIP file from your local hard drive.

5. Open the Jupyter Notebook interface of the SageMaker notebook instance by clicking on the **Open Jupyter** link on the **Notebook Instances** screen. Make sure your instance type is **ml.p2.xlarge**.

6. Once in the notebook instance, navigate to `Applied-Machine-Learning-for-Healthcare-and-Life-Sciences-using-AWS/chapter-6/`

7. Create a directory named data and upload the test folder with NORMAL and PNEUMONIA subfolders inside the data folder on your notebook instance.

After this step, you should have the following directory structure on your notebook instance:

```
medical_image_classification.ipynb
scripts\
        ag_model.py
data\
        test\
                NORMAL\
                PNEUMONIA\
```

You are now ready to begin training a medical image classification model on Amazon SageMaker.

Building, training, and evaluating the model

To train the image classification model, we will be using the **AutoGluon** framework on SageMaker. AutoGluon is an **AutoML** framework that is easy to use and extend. It provides the ability to train models using tabular, imaging, and text data. To learn more about AutoGluon, refer to the guide here: `https://auto.gluon.ai/stable/index.html`.

We will be using a prebuilt AutoGluon container to run training on SageMaker. We will then download the trained model to the notebook instance and run predictions on the model using a test dataset. This demonstrates the **portability** that SageMaker provides by allowing you to train on AWS but run inference locally.

Open the `medical_image_classification.ipynb` notebook and follow the steps in the notebook to complete the exercise.

At the end of the exercise, you will be able to see the confusion matrix shown in *Figure 6.2*:

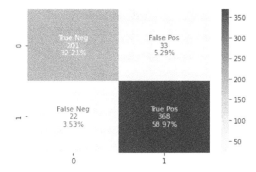

Figure 6.2 – Sample confusion matrix for the medical image classifier

You will also see a ROC curve, as shown in *Figure 6.3*:

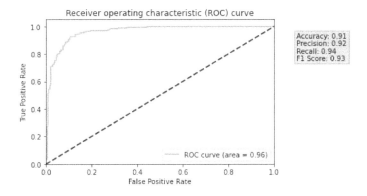

Figure 6.3 – Sample ROC curve for the medical image classifier

This concludes our exercise. Please make sure you stop or delete your SageMaker resources to avoid incurring charges as described at the following link: `https://sagemaker-workshop.com/cleanup/sagemaker.html`.

Summary

In this chapter, we gained an understanding of the medical device industry and the regulatory aspects of the industry designed to ensure the safe usage of these devices among patients. We also looked at the radiology image workflow and the various components involved in the system to make it work. We saw how ML models, when applied to medical devices, can improve the overall health of the population and prevent serious medical events. In the final sections of the chapter, we got an introduction to SageMaker training and went through an implementation exercise to train an ML model to identify a normal chest X-ray compared to one displaying pneumonia.

In *Chapter 7*, *Applying Machine Learning to Genomics*, we will look at how ML technology is transforming the world of clinical research. We will understand the role of genomic sequencing in precision medicine and look at examples that demonstrate why it's important to consider genomic data in clinical diagnosis.

Part 3:
Machine Learning Applications in the Life Sciences Industry

The following chapters will explore use cases for machine learning in life sciences, including how we can work with genes, molecules, and trial data:

7
Applying Machine Learning to Genomics

The **genome** is what defines a living organism. It resides in the cells of the organism and dictates how they perform essential functions. It is also what differentiates one organism from another by providing each organism a set of unique characteristics. Genomes form the complete **deoxyribonucleic acid (DNA)** of an organism. It resides mostly in the nucleus (and in small quantities within the mitochondria) of the cell and provides coded information about the organism by representing it in a sequence of four chemical bases: **thymine (T)**, **cytosine (C)**, **guanine (G)**, and **adenine (A)**. These are arranged in a double helix structure, which is now synonymous with how DNA is represented graphically. The double helix is comprised of two strands of **nucleotides** that bond with each other forming pairs of chemical bases (also known as base pairs), with A pairing with T and G pairing with C. Genomes may also contain sequences of **ribonucleic acid (RNA)**, which is quite common in smaller organisms such as viruses. RNA contains shorter chains of nucleotides than DNA and is synthesized using the information stored in DNA in a process called **transcription**. This results in the formation of proteins and ultimately **phenotypes**, the characteristics of an organism that you can observe.

The human DNA molecule is arranged in structures called **chromosomes** within the cell nucleus. There are 23 pairs of chromosomes (X, Y) in a typical human cell forming a total of 46 chromosomes. The 23rd chromosome differs in males and females, with females having two X chromosomes and males having X and Y each. The **human genome** consists of about 3 billion base pairs of sequences of nucleic acids encoded in the human DNA, which resides in the nucleus of the cell. The **Human Genome Project (HGP)** was a scientific research project kicked off in 1990 and attempted to decode the sequence of the whole human genome, in a process called **genetic sequencing**. It covered 92% of the full human genome sequence as of 2003. Improvements in genomic sequencing technology and lab infrastructure over the years have led to massive gains in our understanding of the human genome. In April 2022, the final remaining 8% of the sequences of the human genome were identified and released. It's fascinating to note that, generally speaking, while no two humans (except identical twins) look the same, it is only 0.1% of human genes that vary between individuals. This is testament to the fact that humans (or *homo sapiens*) are still a very young species compared to other life forms on Earth that pre-date humans by many centuries and, hence, have evolved with many more genetic variations. The 0.1% of genetic variation in humans is the basis of genomic research that attempts to identify how an individual is affected by certain diseases such as diabetes and cancer by comparing the reference human genome with their own. It also helps us diagnose diseases early and discover personalized drugs and therapies that target these variations in individuals.

In this chapter, we will go over the details of genetic sequencing – the process of decoding the information in the genome. We will look at some of the common challenges in processing genomic data generated as a result of genomic sequencing and some common techniques to resolve those challenges. We will also look at how ML can help with the genomic sequencing workflow and our interpretation of genomic data, which is transforming the field of **personalized medicine**. Lastly, we will get introduced to **SageMaker Inference** and build a **Named Entity Recognition** (NER) application to identify genetic entities in a breast cancer report. This helps with downstream processing of genomic testing reports and integrating genomic and clinical entities in **Electronic Health Record** (EHR) systems. Our sections for this chapter are:

- Introducing genomic sequencing
- Challenges with processing genomic data
- Applying ML to genomic workflows
- Introducing SageMaker Inference
- Building a genomic and clinical NER application

Technical requirements

The following are the technical requirements that you need to have in place before building the example implementation at the end of this chapter:

1. Complete the steps to set up the prerequisites for Amazon Sagemaker as described here: `https://docs.aws.amazon.com/sagemaker/latest/dg/gs-set-up.html`

2. Onboard to SageMaker Studio Domain using quick setup as described here: `https://docs.aws.amazon.com/sagemaker/latest/dg/onboard-quick-start.html`

3. Once you are in the Sagemaker studio interface, click on **File | New | Terminal**.

4. Once in the terminal, type the following commands:

   ```
   git clone https://github.com/PacktPublishing/Applied-
   Machine-Learning-for-Healthcare-and-Life-Sciences-using-
   AWS.git
   ```

   You should now see a folder named `Applied-Machine-Learning-for-Healthcare-and-Life-Sciences-using-AWS`.

> **Note**
>
> If you have already cloned the repository in a previous exercise, you should already have this folder. You do not need to do *step 4* again.

5. Familiarize yourself with the SageMaker Studio UI components: `https://docs.aws.amazon.com/sagemaker/latest/dg/studio-ui.html`

6. Create an S3 bucket as described in *Chapter 4*, section *Building a smart medical transcription application on AWS* under *Create an S3 bucket*. If you already have an S3 bucket, you can use that instead of creating a new bucket.

7. Complete the steps to setup Amazon Textract as described here: `https://docs.aws.amazon.com/textract/latest/dg/getting-started.html`

Once you have completed these steps, you should be all set to execute the steps in the example implementation in the last section of this chapter.

Introducing genomic sequencing

Genomic sequencing is the process of decoding the information stored in DNA. As the name suggests, the process attempts to align the sequence of base pairs comprising T, C, G, and A. The combination of these chemicals within the DNA defines the way the cells function at a molecular level. Sequencing the human genome allows us to compare it against a **reference**, which in turn helps us find **mutations** and **variants** in an individual. This is the key to finding how certain diseases affect an individual. It also helps design drugs and therapies that specifically target the mutations of concern. Sequencing the genome of a virus can help us understand it better and develop vaccines that target certain mutations that may be a cause of concern. A recent example of this is the genomic sequencing of the **SARS-CoV-2** virus, which helped us determine multiple mutations and in turn develop modifications to vaccines as the virus evolved.

Over the years, the technology for sequencing has improved significantly. **Next-Generation Sequencing (NGS)**, which involves massively parallel sequencing techniques, has resulted in higher quality and speed in sequencing while reducing the associated cost. It is largely fueled by the advancements in sequencing equipment technology known as sequencers. For example, Illumina, which is the largest manufacturer of sequencers, has sequencers available for a variety of specialized sequencing needs such as **whole genome sequencing (WGS)**, **exome sequencing**, and **RNA sequencing**. This has made genomic sequencing more accessible and practical.

Let us now look at the various stages of whole genome sequencing and how they are categorized.

Categorizing genomic sequencing stages

The whole genome sequencing pipeline involves a range of steps that need seamless integration of hardware and software that work together as a workflow. At a high level, the steps can be categorized under three broad headers:

- **Primary analysis**: Primary analysis consists of a set of steps usually performed by the sequencers. It is at this stage that the analog sample containing genetic matter is converted into a digital data file. The DNA is broken down into smaller molecules to determine the sequence of T, C, G, and A in the genomic sample. This process continues for each part of the DNA molecule and is referred to as a **read**. This process usually results in a **FASTQ file**. The file consists of the sequence string and associated quality score for each read. This is usually referred to as the raw genomic data, which can be further processed to derive insights.

- **Secondary analysis**: The file generated in the primary analysis phase is used as a starting point for the secondary analysis steps. In secondary analysis, the multiple reads are analyzed and filtered to retain only the high-quality ones. This quality control check usually happens on the sequencer. The data from the individual reads are then assembled together to understand the genetic code embedded in the different regions of the DNA. The assembly can be *de novo*, which means it is a scratch assembly with no reference genome to compare it with. When a well-established reference is available, the reads are aligned to the reference genome for comparison. When aligning the sequence, the **depth** of the reads defines how much information about a specific region is considered. The higher the depth during assembly, the more accurate the detection of **variants** is in the genetic code.

- **Tertiary analysis**: The primary task of the tertiary analysis step is detecting variants. This is done using a process called **variant calling**. The variations in the genetic code determine how different the sample is from a reference. The difference can range from a single nucleotide, also known as **single nucleotide polymorphisms** (**SNPs**), to large structural variations. The variants are captured in files known as **variant call files** (**VCFs**) and they are much more manageable in size than the large raw genomic files produced during the primary analysis and the aligned files produced during the secondary analysis. Tertiary analysis is also the basis of downstream research and has all the necessary information to identify new variants and how they affect an individual.

Now that we have an understanding of the whole genome sequencing stages, let us look at how it has evolved over the years.

Looking into the evolution of genomic sequencing

Genomic sequencing aims at reading the genetic code in DNA and aligning it in a sequence for interpretation and analysis. The process originated in the 1970s and was developed by Frederick Sanger, who figured out a way to get the DNA sequence from a sample base by base. He used elaborate chemical reactions and visualization using electrophoresis to achieve this. The process evolved in the 1980s via streamlining sample preparation in the lab and automating visualization of the results on an electropherogram. This technology, also referred to as first-generation sequencing, was used in the Human Genome Project and can handle input sequence lengths of 500-1000 base pairs. While this technology was revolutionary at the time, it was still labor- and time-intensive and the associated costs were quite high. There was a need for higher throughput and lower cost options.

In the mid-2000s, a UK-based company called Solexa (since acquired by Illumina) came up with a technique to amplify a single molecule into many copies across a chip, forming dense clusters or fragments. As technology evolved, the number of these clusters that could be read parallelly into the sequencing machine grew considerably. When you think about sequencing thousands of samples, it is desirable to improve the utilization of each sequencing run by combining multiple samples together. This process, known as multiplexing, allows for the parallel processing of sequencing samples and forms the basis of NGS. This parallel processing technique, also referred to as **second-generation sequencing**, produces reads about 50-500 base pairs in length. It is the most widely used sequencing technique today that is used for rapidly sequencing whole genomes for a variety of clinical and research applications. Organizations such as Edico Genome (since acquired by Illumina) have revolutionized NGS by using **field-programable gate array (FPGA)** based acceleration technology for variant calling.

As the need for whole genome sequencing increased, there was a gap identified while assembling **short-read** sequences. Short-read sequencing that produced 50-500 base pair reads led to incomplete assemblies, sometimes leaving gaps in the sequence. This was especially problematic in de novo sequences where a full reference could not be generated through the short read process. In 2011, **Pacific Biosciences (PacBio)** and **Oxford Nanopore Technologies (ONT)** came up with a **single-molecule real-time (SMRT)** sequencing technology that was capable of generating long reads without amplification. This is now widely known as **third-generation sequencing (TGS)** technology. The most recently released TGS technology from ONT uses nanopores acting as electrodes connected to sensors that can measure the current that flows through it. The current gets disrupted when a molecule passes through the nanopore producing a signal to determine the DNA or the RNA sequence in real time. More recently, in March 2021, a team at Stanford University achieved a time of 5 hours and 2 minutes to sequence a whole human genome, setting the Guinness World Record for the fastest DNA sequencing technique!

As you can see, genomic sequencing has come a long way since the 1970s, when it was introduced, and it is only now that we are beginning to realize its true potential. One of the challenges, however, has been the processing of the vast quantities of data that have been produced during the sequencing process. To address these challenges, advances in cloud computing technology have made scalable storage and compute readily available for genomic workloads. Let us look at some of these challenges and how to address them.

Challenges with processing genomic data

The cost of sequencing a whole human genome has been reduced to less than $1000, making genomic sequencing a part of our regular healthcare. Organizations such as Genomics England in the UK have proposed research initiatives such as the 100,000 Genomes Project to sequence large cohorts to understand how genetic variation is impacting the population. There are similar initiatives underway around the world, in the USA, Australia, and France. A common application of population-scale sequencing is for **genome-wide association studies (GWAS)**. In these studies, scientists look at the entire genome of a large population searching for small variations to help identify particular diseases or traits. While this is extremely promising, it does introduce the technical problem of processing all this data that gets generated as part of the genomic sequencing process, especially when we are talking about sequencing hundreds of thousands of people.

Let us now look at some of the challenges that organizations have to face when processing genomic data.

Storage volume

The data generated by a single whole genome sequencing of a human can generate about 200 GB of data. Some estimates suggest that genomic sequencing driven by research will generate anywhere between 2 and 40 exabytes of data within the next decade. To add to the complexity, the files range from small- to medium-sized files of a few MBs to large files of a few GBs. The data comes in different formats and needs to be arranged in groups that can be partitioned by runs or samples. This is a surge of genomic big data and is quickly filling up on-premises data centers where storage volumes are limited by data center capacities. One item of good news about genomic data is that as the data progresses from its raw format to formats where it can be used to derive meaning, its size condenses. This allows you to tier your storage volumes that store genomic data. This means customers can choose a high-throughput storage tier for the most condensed processed genomic files that consume less space and move the less utilized raw files to a low storage tier or even archive them. This helps manage storage costs and volumes.

One ideal option to store genomic data on AWS is Amazon S3, which is an object store. It is highly suited for genomic data as it is scalable and provides multiple storage tiers and the ability to move files between storage tiers automatically. In addition, AWS provides multiple storage options outside of S3 for a variety of use cases that support genomic workflows.

To learn more about AWS storage services, you can look at the following link: `https://aws.amazon.com/products/storage/`.

Sharing, access control, and privacy

Genomic data contains a treasure trove of information about a population. If hackers get access to genetic data about millions of people, they would be able to derive information about their heritage, their healthcare history, and even their tolerance or allergies to certain food types. That's a lot of personal information at risk. Direct-to-consumer genomic organizations such as 23andMe and Helix are examples of organizations that collect and maintain such information. It is therefore vital for organizations that collect this data to manage it in the most secure manner. However, genomic research generally involves multiple parties and requires sharing and collaboration.

An ideal genomic data platform should be securely managed with strict access control policies. The files should ideally be encrypted so they cannot be deciphered by hackers in the event of a breach. The platform should also support selective sharing of files with external collaborators and allow selectively receiving files from them. Doing this on a large scale can be challenging. Hence, organizations such as Illumina provide managed storage options for their genomic sequencing platforms that support access control and security out of the box for the data generated from their sequencing instruments. Moreover, AWS provides robust encryption options, monitoring, access control via IAM policies, and account-level isolation for external parties. You can use these features to create the desired platform for storing and sharing your genomic datasets.

To learn more about AWS security options, you can look at the following link: `https://aws.amazon.com/products/security/`.

The Registry of Open Data is another platform built by AWS to facilitate sharing of genomics datasets and support collaboration among genomic researchers. It helps with the discovery of unique datasets available via AWS resources. You can find a list of open datasets hosted in the registry at the following link: `https://registry.opendata.aws`.

Compute

Genomic workloads can be treated as massively parallel workflows. They benefit greatly from parallelization. Moreover, the size and pattern of the compute workloads depend on the stage of the processing. This creates a need for a homogeneous compute environment that supports a variety of instance types. For example, some processes might be memory bound and need high memory instances, and some might be CPU bound and might need large CPU instances. Others might be GPU bound. Moreover, the size of these instances may also need to vary as the data moves from one stage to another. If we have to summarize three key requirements of an ideal genomic compute environment, they would be the following:

- Able to support multiple instance types, such as CPU, GPU, and memory bound
- Able to scale horizontally (distributed) and vertically (larger size)
- Able to orchestrate jobs across different stages of the genomic workflow

On AWS, we have a variety of options that meet the preceding requirements. You can build and manage your own compute environment using EC2 instances or use a managed compute environment to run your compute jobs. To know more about compute options on AWS, you can look at the following whitepaper: `https://docs.aws.amazon.com/whitepapers/latest/aws-overview/compute-services.html`.

Additionally, AWS offers a purpose-built workflow tool for genomic data processing called the Genomics CLI. It automatically sets up the underlying cloud infrastructure on AWS and works with industry-standard languages such as **Workflow Description Language** (**WDL**). To know more about the Genomics CLI, you can refer to the following link: `https://aws.amazon.com/genomics-cli/`.

Interpretation and analysis

With the rapid pace of large-scale sequencing, we are being outpaced in our ability to interpret and draw meaningful insights from genomic data compared to the volumes we are sequencing. This introduces a need to use advanced analytical and ML capabilities to interpret genomic data. For example, correlating genotypes with phenotypes needs a query environment that supports distributed query techniques using frameworks such as Spark. Researchers also need to query reference databases to search through thousands of known variants to determine the properties of a newly discovered variant. Another important analysis is to correlate genomic data with clinical data to understand how a patient's medical characteristics such as medications and treatments impact their genetic profile and vice versa. The discovery of new therapies and drugs usually involves carrying out such analysis at large scales, which the genomic interpretation platforms need to support. Analytical services from AWS provide easy options for the interpretation of genomic data. For example, Athena, which is a serverless ad hoc query environment from AWS, helps you correlate genomic and clinical data without the need for spinning up large servers and works directly on data stored on S3. For more complex analysis, AWS has services such as **Elastic Map Reduce** (**EMR**) and **Glue** that support distributed frameworks such as Spark. To know more about analytics services from AWS, you can look at the following link: `https://aws.amazon.com/big-data/datalakes-and-analytics/`.

In addition to the preceding resources, AWS also published an end-to-end genomic data platform whitepaper that can be accessed here: `https://docs.aws.amazon.com/whitepapers/latest/genomics-data-transfer-analytics-and-machine-learning/genomics-data-transfer-analytics-and-machine-learning.pdf`.

Let us now dive into the details of how ML can help optimize genomic workflows.

Applying ML to genomic workflows

ML has become an important technology, applied throughout the genomic sequencing workflow and the interpretation of genomic data in general. ML plays a role in data processing, deriving insights, and running searches, which are all important applications in genomics.

One of the primary drivers of ML in genomic workflows is the sheer volume of genomic data to analyze. As you may recall, ML relies on pattern recognition in unseen data that has been learned from a previous subset of data. This process is much more computationally efficient than applying complex rules on large genomic datasets. Another driver is that the field of ML has evolved, and computational resources such as GPUs have become more accessible to researchers performing genomic research. Proven techniques can now produce accurate models on different modalities of genomic information. Unsupervised algorithms such as **principal component analysis** (**PCA**) and **clustering** help pre-process genomic datasets with a large number of dimensions. Let us now look at different ways in which ML is being applied to genomics:

- **Labs and sequencers**: Sequencing labs have connectivity to the internet that allows them to take advantage of using ML models for predictive maintenance of lab instruments. Moreover, models deployed directly on the instruments can alert users about low volumes of reagents and other consumables in the machine that could disrupt sequencing. In addition to maintenance, another application of ML on genomic sequencers is in the process of quality checks. As described earlier, sequencers generate large volumes of short reads, which may or may not be of high quality. Using ML models that run on the sequencers, we can filter out low-quality reads right on the machines, thereby saving downstream processing time and storage.

- **Disease detection and diagnosis**: Disease detection and diagnosis is a common area of application of ML in genomics. Researchers are utilizing ML algorithms to detect attributes of certain types of diseases from gene expression data. This technique is proven to be very useful in diseases that progress over a period of time such as cancer and diabetes. The key to treating these diseases is early diagnosis. Using ML algorithms, researchers can now identify unique patterns that correlate to diseases much earlier in the process compared to conventional testing techniques. For example, imaging studies to detect cancerous tumors can diagnose a positive case of cancer only when the tumor has already formed and is visible. Compared to this, a genomic test for cancer diagnosis can find traces of cancer DNA in the bloodstream of patients way before the tumor has been formed. These approaches are leading to breakthroughs in the field of cancer treatment and may one day lead to a cancer-free world.

- **Discovery of drugs and therapies**: Variants in a person's genetic data are helping with the design of new drugs and therapies that target those variants. ML models trained on known variants can determine whether detected variants in a patient are new and even classify them based on their properties. This helps with understanding the variant and how it would react to medications. Moreover, by combining genomic and proteomic (a sequence of proteins) data, we can identify genetic biomarkers that develop certain types of proteins and even predict their properties, such as toxicity and solubility, and even how those proteins are structured. By sequencing genomes of viruses, we can find unique characteristics in the virus's genetic code, which in turn helps us design vaccines that affect the targeted regions of the genetic code of the virus, making them harmless to humans.

- **Knowledge mining and searches**: Another common area of application of ML in genomics is the ability to mine and search all the existing information about genetics in circulation today. This information can exist in curated databases, research papers, or on government websites. Mining all this information and searching through it is extremely difficult. We can now have large genomic language models that have been pre-trained on a variety of corpuses containing genetic information. By applying these pre-trained models to genomic text, we are able to get an out-of-the-box capability for a variety of downstream ML tasks such as classification and **NER**. We can also detect similarities between different genomic documents and generate topics of interest from documents. All these techniques help with indexing, searching, and mining genomic information. We will be using such a pre-trained model for an NER task in the last section of this chapter.

Now that we have an understanding of genomics and the associated workflows, let us get a deeper look into SageMaker Inference and use it to build an application for extracting genomic and clinical entities.

Introducing Amazon SageMaker Inference

In *Chapter 6*, you learned about SageMaker training and we dove into the details of the various options available to you when you train a model on SageMaker. Just like training, SageMaker provides a range of options when it comes to deploying models and generating predictions. These are available to you via the SageMaker Inference component. It's important to note that the training and inference portions of SageMaker are decoupled from one another. This allows you to choose SageMaker for training, inference, or both. These operations are available to you via the AWS SDK as well as a dedicated SageMaker Python SDK. For more details on the SageMaker Python SDK, see the following link: `https://sagemaker.readthedocs.io/en/stable/`.

Let us now look at the details of options available to deploy models using SageMaker.

Understanding real-time endpoint options on SageMaker

In this option, SageMaker provides a persistent endpoint where the model is hosted and available for predictions 24/7. This option allows you to generate predictions with low latency in real time. SageMaker manages the backend infrastructure to host the model and make it available via a REST API for predictions. It attaches an EBS volume to the underlying instance to store model artifacts. The inference environment in SageMaker is containerized. You have the option to use one of the SageMaker supported Docker images or provide your own Docker image and inference code. SageMaker also manages scaling the inference endpoint on multiple instances via autoscaling if needed. To know more about autoscaling of models on SageMaker, see the following link: `https://docs.aws.amazon.com/sagemaker/latest/dg/endpoint-auto-scaling.html`.

Since real-time endpoints are persistent, it is important to use them sparingly and delete them when not in use to avoid incurring charges. It is also important to consider deployment architectures that can merge multiple models behind one endpoint if needed. When hosting a model on a SageMaker endpoint, you have the following options to consider:

- **Single model**: This is the simplest option, where a single model is hosted behind an endpoint. You need to provide the S3 path to the model artifacts and the Docker image URI of the host container. In addition, common parameters such as region and role are needed to make sure you have access to the backend AWS resources.
- **Multiple models in one container**: You can use a SageMaker multi-model endpoint to host multiple models behind a single endpoint. This allows you to deploy models cost-effectively on managed infrastructure. The models are hosted in a container that supports multiple models and shares underlying resources such as memory. It is therefore recommended that if your model has high latency and throughput requirements, you should deploy it on a separate endpoint to avoid any bottlenecks due to sharing of resources. SageMaker frameworks such as `TensorFlow`, `scikit-learn`, `PyTorch`, and `MxNet` support multi-model endpoints. So do built-in algorithms such as Linear Learner, XGBoost, and k-nearest neighbors. You can

also use the SageMaker Inference tools to build a custom container that supports multi-model endpoints. To learn more about this option, you can refer to the following link: `https://docs.aws.amazon.com/sagemaker/latest/dg/build-multi-model-build-container.html`.

When you host multiple models on a SageMaker endpoint, SageMaker intelligently loads and unloads the model into memory based on the invocation traffic. Instead of downloading all the model artifacts from S3 when the endpoint is created, it only downloads them as needed to preserve precious resources. Once the model is loaded, SageMaker serves responses from that model. SageMaker tracks memory utilization of the instance and routes traffic to other instances if available. It also unloads the model from memory and saves it on the instance storage volume if the model is unused and another model needs to be loaded. It caches the frequently used models on the instance storage volume, so it doesn't need to download them from S3 every time it's invoked.

- **Multiple models in different containers**: In some cases, you may have a need to generate inference in sequence from multiple models to get the final value of the prediction. In this case, the output of one model is fed into the next. To support this pattern of inference, SageMaker provides a multi-container endpoint. With the SageMaker multi-container endpoint, you can host multiple models that use different frameworks and algorithms behind a single endpoint, hence improving the utilization of your endpoints and optimizing costs. Moreover, you do not need to invoke the models sequentially. You have the option to directly call the individual containers in the multi-container endpoint. It allows you to host up to 15 models in a single endpoint and you can invoke the required container by providing the container name as an invocation parameter.

- **Host models with preprocessing logic**: With SageMaker Inference, you have the option to host multiple containers as an inference pipeline behind a single endpoint. This feature is useful when you want to include some preprocessing logic that transforms the input data before it is sent to the model for prediction. This is needed to ensure that the data used for predictions is in the same format as the data that was used during training. Preprocessing or postprocessing data may have different underlying infrastructure requirements than generating predictions from the model. For example, SageMaker provides Spark and `scikit-learn` containers within SageMaker processing to transform features. Once you determine the sequence of steps in your inference code, you can stitch them together by providing the individual container Docker URI while deploying the inference pipeline.

Now that we have an understanding of real-time inference endpoint options on SageMaker, let us look at other types of inference architectures that SageMaker supports.

Understanding Serverless Inference on SageMaker

SageMaker Serverless Inference allows you to invoke models in real time without persistently hosting them on an endpoint. This is particularly useful for models that have irregular or unpredictable traffic patterns. Keeping a real-time endpoint idle is cost prohibitive so Serverless Inference scales down the number of instances to zero when the model is not invoked, thereby saving costs. It is also important to note that there is a delay in invocation response for the first time when the model is loaded on the instance. This is known as a cold start. It is therefore recommended that you use Serverless Inference for workflows that can tolerate cold start periods.

SageMaker Serverless Inference automatically scales instances behind the endpoint based on the invocation traffic patterns. It can work with one of the SageMaker framework or algorithm containers or you can create a custom container and use it with Serverless Inference. You have the option to configure the memory footprint of the SageMaker serverless endpoint ranging from 1 GB to 6 GB of RAM. It then auto-assigns the compute resources based on the memory size. You also have a limit of 200 concurrent invocations per endpoint so it's important to consider latency and throughput in your model invocation traffic before considering Serverless Inference.

Understanding Asynchronous Inference on SageMaker

A SageMaker asynchronous endpoint executes model inference by queuing up the requests and processing them asynchronously from the queue. This option is well suited for scenarios where you need near-real-time latency in your predictions and your payload size is really large (up to 1 GB), which causes long processing times. Unlike a real-time persistent endpoint, SageMaker Asynchronous Inference spins up infrastructure to support inference requests and scales it back to zero when there are no requests, thereby saving costs. It uses the **simple notification service** (**SNS**) to send notifications to the users about the status of the Asynchronous Inference. To use SageMaker Asynchronous Inference, you define the S3 path where the output of the model will be stored and provide the SNS topic that notifies the user of the success or error. Once you invoke the asynchronous endpoint, you receive a notification from SNS on your medium of subscription, such as email or SMS. You can then download the output file from the S3 location you provided while creating the Asynchronous Inference configuration.

Understanding batch transform on SageMaker

Batch transform allows you to run predictions on the entire dataset by creating a transformation job. The job runs asynchronously and spins up the required infrastructure, downloads the model, and runs the prediction on each input data point in the prediction dataset. You can split the input data into multiple files and distribute the inference processing across multiple instances to distribute the data and optimize performance. You can also define a mini-batch of records to load at once from the files for inference. Once the inference job completes, SageMaker creates a corresponding output file with the extension .out that stores the output of the predictions. One of the common tasks after generating predictions on large datasets is to associate the prediction output with the original features in the prediction data that may have been excluded from the prediction data. For example, an ID column that may be excluded from the prediction data when generating predictions may need to be re-associated with the predicted value. SageMaker batch transform provides a parameter called `join_source` to allow to you achieve this behavior. You also have the option of choosing the `InputFilter` parameter and `OutputFilter` parameter to specify a portion of the input to pass to the model and the portion of the output of the transform job to include in the output file.

As you can see, SageMaker Inference is quite versatile in the options it provides for hosting models. Choosing the best option usually depends on the use case, data size, invocation traffic patterns, and latency requirements. To learn more about SageMaker inference, you can refer to the following link: `https://docs.aws.amazon.com/sagemaker/latest/dg/deploy-model.html`.

In the next section, we deploy a single pre-trained model on a SageMaker real-time endpoint for detecting genomic entities from unstructured text extracted from a genomic testing report.

Building a genomic and clinical NER application

In *Chapter 4*, you saw how you can extract clinical entities from audio transcripts of a patient visit summary using Comprehend Medical. Sometimes, we may have the need to extend the detection of named entities beyond the set of clinical entities that Comprehend Medical detects out of the box. For example, genomic entities can be detected from genomic testing reports and be combined with clinical entities from Comprehend Medical to get a better understanding of both categories of entities in the report, resulting in better interpretation. This also helps automate the clinical genomic reporting pipeline by automating the process of extracting meaningful information hidden in the testing report. To better understand the application, look at the following diagram.

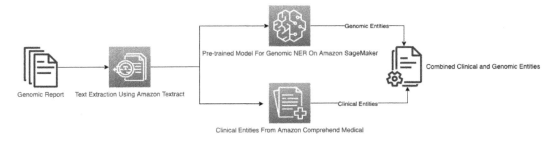

Figure 7.1 – Workflow for the NER application

The purpose of the application is to create an automated workflow to extract genetic and clinical entities from a genomic testing report. As shown in the preceding diagram, we will be using three key AWS services to build our application:

- **Amazon Textract**: We will use Textract to analyze the genomic testing report in PDF format and get the line-by-line text on each page of the report.

- **Amazon SageMaker**: We will use a SageMaker real-time model endpoint hosting a pre-trained model from the Hugging Face Model Hub to detect genetic entities from unstructured text. We will send lines of text from the genomic testing report to this model to detect genetic entities.

- **Amazon Comprehend Medical**: We will use Amazon Comprehend Medical to extract clinical entities from the same text in the report.

The final result will have all the genomic as well as clinical entities detected from the report. Before you begin, make sure you have access to SageMaker Studio and have set up Amazon Textract and S3 as described in the prerequisites section.

Acquiring the genomic test report

We will begin by downloading a sample test report from **The Cancer Genome Atlas (TCGA)** portal. This will be our source of the text on which we will run NER models.

> **Note**
>
> This exercise uses the Jupyter notebook that we downloaded in the *Technical requirements* section. The notebook has instructions and documentation about the steps in the exercise. Follow these steps to set up the notebook.

1. To download the sample report, visit the following URL: `https://portal.gdc.cancer.gov/repository?filters=%7B%22op%22%3A%22and%22%2C%22content%22%3A%5B%7B%22op%22%3A%22in%22%2C%22content%22%3A%7B%22field%22%3A%22files.data_format%22%2C%22value%22%3A%5B%22pdf%22%5D%7D%7D%5D%7D`.

   The URL opens up the TCGA portal and applies the appropriate filters to show you a list of PDF reports. These are redacted pathology reports for breast cancer testing and consist of multiple pages.

2. You can pick any report from this list to use in this application. To make it easy for you to follow along, let's pick the file `ER0197 ER-ACOF Clinical Report_Redacted_Sanitized.pdf`. Click on the hyperlink for this file to open the download page. Click on the download link. There is a search field on the left where you can type `ER0197` to help you find the file easily.

3. Unzip/untar the downloaded file. This will create a folder with the name starting with `gdc_download_`. There will also be another folder within that whose name will look like a hash key. Inside this folder, you will see a PDF file. This is the source PDF file that we will use for our models.

> **Note**
>
> If you are using a Windows-based computer, you can use WinZip to untar the `tar` file. See more details can be found here: `https://www.winzip.com/en/learn/file-formats/tar/`.

4. Open the PDF file and examine its contents. Once you are ready, upload the PDF file to the S3 bucket you created as part of the prerequisites.

You are now ready to run the notebook `genetic_ner.ipynb` to complete the exercise. However, before you do that, let us get a better understanding of the pre-trained model that we use in this exercise.

Understanding the pre-trained genetic entity detection model

As described earlier, we will be using a pre-trained NER model to detect genetic entities. The model we will use is available as a Hugging Face transformer. Hugging Face transformers provide an easy way to fine-tune and use pre-trained models for a variety of tasks involving text, tabular data, images, or multimodal datasets. Hugging Face maintains these models in a central repository known as the Hugging Face Hub. You can reference these models directly from the Hub and use them in your workflows. To learn more about Hugging Face transformers, you can visit the following link: `https://huggingface.co/docs/transformers/index`.

The model we will be using today is available on the Hugging Face Hub at the following link: `https://huggingface.co/alvaroalon2/biobert_genetic_ner`.

The model we will use is known as `biobert_genetic_ner`. The model is a fine-tuned version of the **BioBERT** model for NER tasks. It has been fine-tuned on certain datasets such as **Bio Creative II Gene Mention Recognition** (BC2GM). To learn more about the model, visit the Hugging Face Hub link: `https://huggingface.co/alvaroalon2/biobert_genetic_ner`.

Although not required to complete this exercise, it is also a good idea to familiarize yourself with transformer-based architecture for language understanding. It uses deep bidirectional transformers to understand unlabeled text and has achieved state-of-the-art performance on a variety of NLP tasks. You can learn more about it at the following link: `https://arxiv.org/abs/1810.04805`.

Once you are ready, open the notebook genetic_ner.ipynb in SageMaker Studio by navigating to `Applied-Machine-Learning-for-Healthcare-and-Life-Sciences-using-AWS/chapter-7/`. The notebook has instructions to complete the exercise. At the end of the exercise, you will see the output from the model that recognizes genomic and clinical entities from the test report. Here is a snapshot of what the output will look like:

```
- Her2neu
    Type: GENETIC

- immunoperoxidase panel
    Type: TEST_NAME
    Category: TEST_TREATMENT_PROCEDURE

- immunoperoxidase
```

```
    Type: GENETIC

- cleared
  Type: DX_NAME
  Category: MEDICAL_CONDITION
  Traits:
   - NEGATION
```

Let's review what we've learned in this chapter.

Summary

In this chapter, we went into the basics of genomics. We defined key terminology related to genetics and understood how genomes influence our day-to-day functions. We also learned about genomic sequencing, the process of decoding genomic information from our DNA. We then learned about the challenges that organizations face when processing genomic data at scale and some techniques they can use to mitigate those challenges. Lastly, we learned about the inference options of SageMaker and hosted a pre-trained model on a SageMaker real-time endpoint. We combined this model with clinical entity recognition from Comprehend Medical to extract genetic and clinical entities from a genomic testing report.

In *Chapter 8*, *Applying Machine Learning to Molecular Data*, we will look at how ML is transforming the world of molecular property prediction. We will see how ML-based approaches are allowing scientists to discover unique drugs faster and with better outcomes for patients.

Further reading

- Understanding a genome: https://www.genome.gov/genetics-glossary/Genome
- DNA sequencing fact sheet: https://www.genome.gov/about-genomics/fact-sheets/DNA-Sequencing-Fact-Sheet
- Definition of DNA: https://www.britannica.com/science/DNA
- Next-Generation Sequencing: https://www.ncbi.nlm.nih.gov/pmc/articles/PMC3841808/
- Genomic data processing overview: https://gdc.cancer.gov/about-data/gdc-data-processing/genomic-data-processing

Applying Machine Learning to Molecular Data

Molecular data contains information about the chemical structures of different molecules. Extrapolating this data for analysis helps us determine important properties of a chemical at a molecular level. These properties are key in the discovery of new therapies and drugs. For example, small molecules with specific atomic structures can combine with other molecules to form compounds. These compounds may then become compounds of interest or **candidate compounds** if they are beneficial in treating diseases. This is done by looking at the interaction of the compound with a protein or gene involved in a disease and understanding how this interaction affects the underlying protein or gene. If the interaction helps to modulate the function or behavior of the gene or the protein, we can say that we have found a **biological target** for our compound. This process, also known as **target discovery**, is a key step in discovering new drugs. In some cases, instead of scanning known compounds, researchers can discover drugs by designing compound structures that have an affinity for the target. By bonding with the target, the compounds can change the function of the disease-causing target to render them harmless to humans. This process is known as **drug design**. Once we identify the compound for a target, the process moves toward determining the compound's properties such as its **solubility** or **toxicity**. This helps us to understand the positive or negative effects it could have on humans and also the exact quantities that need to be administered to make sure it has the desired effects on the target.

However, all this is easier said than done. There are billions of compounds to consider and a growing list of diseases to correlate them with. This introduces a search problem at an insurmountable scale, with the search sometimes taking several months to complete. Traditional computing techniques rely on the brute-force method of target identification, which essentially involves going through a whole list of probable compounds and evaluating them based on their effects on diseases, based on trial and error. This process is not scalable and introduces bottlenecks. As a result, researchers now rely on techniques such as ML that can infer patterns in the chemical composition of known compounds and use that to predict the interactive effects of novel compounds on targets not yet studied or known.

In this chapter, we will look at the various steps in drug discovery. We will see how molecular data is stored and processed during the drug discovery process and also understand different applications of

ML in the areas of drug design and discovery. We will look at a feature of SageMaker that allows you to bring your own custom container for training, allowing you to package specific software that can work with molecular data. Lastly, we will build an ML model to predict different molecular properties.

We will cover the following topics:

- Understanding molecular data
- Introducing drug discovery and design
- Applying ML to molecular data
- Introducing custom containers in SageMaker
- Building a molecular property prediction model on SageMaker

Technical requirements

The following are the technical requirements that you need to complete before building the example implementation at the end of this chapter:

1. Complete the steps to set up the prerequisites for Amazon SageMaker as described here: https://docs.aws.amazon.com/sagemaker/latest/dg/gs-set-up.html.

2. Create a SageMaker notebook instance by following the steps in the following guide: https://docs.aws.amazon.com/sagemaker/latest/dg/howitworks-create-ws.html.

3. Create an S3 bucket as described in *Chapter 4*, in the *Building a smart medical transcription application on AWS* section under *Create an S3 bucket*. If you already have an S3 bucket, you can use that instead of creating a new bucket.

4. Open the Jupyter notebook interface of the SageMaker notebook instance by clicking on the **Open Jupyter** link on the Notebook Instances screen.

5. In the top-right corner of the Jupyter notebook, click on **New** and then **Terminal**.

6. Type the following commands on the terminal screen:

```
$cd SageMaker
$git clone https://github.com/PacktPublishing/Applied-
Machine-Learning-for-Healthcare-and-Life-Sciences-using-
AWS.git
```

You should now see a folder named Applied-Machine-Learning-for-Healthcare-and-Life-Sciences-using-AWS.

> **Note**
>
> If you have already cloned the repository in a previous exercise, you should already have this folder. You do not need to do *step 6* again.

Once you have completed these steps, you should be all set to execute the steps in the example implementation in the last section of this chapter.

Understanding molecular data

Having a good understanding of molecular properties and structures is extremely critical to determine how they react with each other. These reactions lead to the discovery of new molecules that lead to drug development. **Pharmacology** is the branch of science that studies such reactions and their impact on the body. Pharmacologists do this by reviewing molecular data stored in a variety of formats. At a very high level, molecules can be divided into two categories, small and large molecules. The distinction between them is not just because of their size. Let's look at them in more detail.

Small molecules

Small molecules have been the basis of drug development for a very long time. They weigh less than 900 Dalton (Da) (1 Da is equal to $1.66053904 \times 10^{-24}$ grams) and account for more than 90% of drugs on the market today. Drugs based on small molecules are mostly developed through chemical synthesis. Due to their small size, they are easily absorbed into the bloodstream and can reach their biological targets through cell membranes to induce a biological response and can be administered orally as tablets. There are several small molecule databases, such as the **small molecule pathway database** (**SMDB**), which contains comprehensive information on more than 600 small molecule metabolic pathways found in humans. Their structures can be determined via well-known techniques such as **spectroscopy** and **X-ray crystallography**, which can determine the mean positions of the atoms and the chemical bonds allowing us to infer the structure of the molecule. A common way to represent the structure of the molecule is by using **simplified molecular-input line-entry system** (**SMILES**) notation. SMILES allows you to represent molecular structures using a string and is a common feature in small molecule data. SMILES is also useful to create a graph representation of a molecular structure, with atoms representing the nodes and the bonds between them representing the edges of the graph. This type of graph representation of molecular structure allows us to create a chemical graph of the compound and feed that as a feature for downstream analysis of the compound, using graph query language such as **Gremlin** with data stored in **graph databases** such as **Neo4j**.

Large molecules

Large molecules or macromolecules can be anywhere between 3,000 and 150,000 Daltons in size. They include biologics such as proteins that can interact with other proteins and create a therapeutic effect. Large molecules are more complex than their small counterparts and usually need to be injected into the body instead of taken orally. Drug development with large molecules involves engineering cells to bind with site-specific biological targets to induce a more localized response, instead of interfering with the functions of the overall body. This has made biologics more popular for site-specific targeted therapies for diseases such as cancer, where a tumor may be localized to a few areas of the body. The structure of large molecules such as proteins can be determined using technologies such as crystallography and **cryo-electron microscopy** (**cryo-EM**), which uses electron beams to capture the image of radiation-sensitive samples in an electron microscope. There are several databases of known protein structures, such as the **protein databank** (**PDB**), that researchers can use to study the 3D representations of proteins. More recently, our effort to understand proteins took a huge leap with the announcement of **Alphafold** by DeepMind in 2021. It is an AI algorithm that can accurately predict the structure of a protein from a sequence of amino acids. This has led to a slew of research targeted toward large molecule therapies that involve **structure-based drug design**.

With this understanding of molecular data types and their role in drug development, let us now dive into the details of the drug discovery process by looking at its various stages.

Introducing drug discovery and design

The process of drug discovery and design centers around the identification of biological targets. A disease may demonstrate multiple clinical characteristics, and the goal of the drug compound is to modulate the behavior of the biological entity (such as proteins and genes) that can change the clinical behavior of the disease. This is referred to as the target being **druggable**. The compounds do this by binding to the target and altering its molecular structure. As a result, understanding the physical structure of the molecule is extremely critical in the drug discovery and design process. **Cheminformatics** is a branch of science that studies the physical structure as well as the properties of molecules, such as **absorption, distribution, metabolism, and extraction** (**ADME**). Targets are screened against millions of compounds based on these properties to find the candidates that can be taken through the steps of clinical research and clinical trials. This process is known as **target-based drug discovery** (**TBDD**) and has the advantage of researchers being able to target the specific properties of a compound to make it suitable for a target. A contrasting approach to TBDD is **phenotypic screening**. In this approach, having prior knowledge of the biological target is not necessary. Instead, the process relies on monitoring changes to a specific **phenotype**. This provides a more natural environment for a drug to interact with a biological molecule to track how it alters the phenotype. However, this method is more difficult as the characteristics of the target are not known. There are multiple methods for target discovery, ranging from simple techniques such as literature survey and competitive analysis to more complex analysis techniques, involving simulations with genomic and proteomic datasets and **chromatographic** approaches. The interpretation of data generated from these processes requires specialized technology, including high-performance computing clusters with GPUs, image-processing pipelines, and search applications.

In the case of small molecule drug discovery, once a target is identified, it needs to be validated. This is the phase when the target is scrutinized to demonstrate that it does play a critical role in a disease. It is also the step where the drug's effects on the target are determined. The in-depth analysis during the validation phase involves the creation of biological and compound screening **assays**. Assays are procedures defined to measure the biological changes as a result of interaction between the compound and the target. The assay should be reproducible and robust to false positives and false negatives by having a balance between selectivity and sensitivity. This helps shortlist a few compounds that are relevant as they demonstrate the desired changes in the drug target. These compounds are referred to as a hit **molecule**. There could be several hit molecules for a target. The process of screening multiple potential compound libraries for hits is known as **high-throughput screening (HTS)**.

HTS involves scanning through thousands of potential compounds that have the desired effects on a target. HTS uses automation and robotics to speed up this search process by quickly performing millions of assays across entire compound libraries. The assays used in HTS vary in complexity and strategy. The type of assay used in HTS depends on multiple factors, such as the biology of the target, the automation used in the labs, the equipment utilized in the assay, and the infrastructure to support the assay. For example, assays can measure the affinity of a compound to a target (**biochemical assays**) or use fluorescence-based detection methods to measure the changes in the target (luminescent assays). We are now seeing the increased use of virtual screening, which involves creating simulation models for assays that can screen a large number of compounds. These virtual screening techniques are powered by ML models and have dramatically increased the speed with which new targets can be matched up to potential compounds.

The steps following HTS largely consist of tasks to further refine the hits for potency and test out their **efficacy**. It involves analyzing ADME properties along with structure analysis using **crystallography** and **molecular modeling**. This is typically known as **lead optimization** and results in a candidate that can be taken through the clinical research phase for drug development.

In contrast to small molecules, therapies involving large molecules are known to target more serious illnesses, such as cancers. As large molecules have a more complicated structure, the process of developing therapies with them involves specialized processes and non-standard equipment. It is also a more costly affair with a larger upfront investment and comes with considerable risks. However, this can be a competitive advantage to a pharmaceutical organization because the steps are not as easy to reproduce by other pharmaceutical organizations. It is also known that cancers can develop resistance to small molecules over a period of time. Such discoveries have led to the formation of a variety of start-ups doing cutting-edge research in **CAR T-cell therapy** and **gene-based therapy**. These efforts have made biologics more common in recent years.

While drug discovery has gained momentum with the advancements in recent technology, it is still a costly and time-consuming affair. In contrast, researchers are now able to design a drug based on their knowledge of the biological target. Let's understand this better.

Understanding structure-based drug design

The process of drug design revolves around creating molecules that have a particular affinity for a target in question and can bind to it structurally. This is done by designing molecules complementary in design to the target molecule. The structure and the knowledge of the target are extremely important in the design process. It involves computer simulation and modeling to synthesize and develop the best candidate compounds that have an affinity for the target. This is largely categorized as **computer-aided drug design** (CADD) and has become a cost-effective option for designing promising drug candidates in preference to In Vitro methods. With increasing knowledge about the structure of various proteins, more potential drug targets are being identified.

The structure of a target is determined using techniques such as **X-ray crystallography** or **homology modeling**. You can also use more advanced techniques such as **cryo-EM** or AI algorithms such as **Alphafold** to ascertain the structure of the target. Once the target is identified and its structure is determined, the process aims at determining the possible binding sites on the target. Compounds are then scored based on their ability to bind with the target. The high-scoring ones are developed into assays. This target-compound structure is further optimized with multiple parameters of the compound that determine its effectiveness as a drug. These include parameters such as the stability and toxicity of the compounds. The process is repeated multiple times and aims to determine the structure and optimize the properties of the compound-target complex. After several rounds of optimization, it is highly likely that we will have a compound that will be specific to the target in question.

At the end of the day, drug discovery and design are both iterative processes that try to find the most probable compound that has an affinity to a target. The discovery process does it via search and scan whereas the design process achieves it via optimization of different parameters. Both of them have largely been modernized by the use of ML, and their implementation has become much more common in recent years. Let us now look at various applications of ML on molecular data that have led to advances in drug discovery and design.

Applying ML to molecular data

The discovery of a drug requires a trial-and-error method that involves scanning large libraries of small molecules and proteins using high-performance computing environments. ML can speed up the process by predicting a variety of properties of the molecules and proteins, such as their toxicity and binding affinity. This reduces the search space, thereby allowing scientists to speed up the process. In addition, drug manufacturers are looking at ways to customize drugs to an individual's biomarkers (also known as precision medicine). ML can speed up these processes by correlating molecular properties to clinical outcomes, which helps in detecting biomarkers from a variety of datasets such as biomedical images, protein sequences, and clinical information. Let us look at a few applications of ML on molecular data.

Molecular reaction prediction

One of the most common applications of ML in drug discovery is the prediction of how two molecules would react with each other. For instance, how would a protein react with a drug compound? These models can be pre-trained on existing data, such as the **drug protein interaction** (DPI) database, and can be used to predict future unknown reactions in a simulated fashion. We can apply such models for **repurposing drugs**, a method that encourages using existing drug candidates that were shortlisted for some clinical condition but instead uses them for another condition. Using these reaction simulations, you can get a better idea of how the candidate would react to the new target and, hence, determine whether further testing is needed. This method is also useful for target identification. A target provides binding sites (also known as druggable sites) for candidate molecules to dock. In a typical scenario, the identification of new targets requires the design of complicated assays for multiple proteins.

Instead, an in silco approach reduces the time to discover new targets by reducing the number of experiments. A **drug target interaction (DTI)** model can predict whether a protein and a molecule would react. Another application of this is for biologics design. Such therapies depend on macromolecule reactions such as **protein-protein interactions (PPIs)**. These models attempt to understand the pathogenicity of the host protein and how it interacts with other proteins. An example implementation of this is in the process of vaccine development for known viruses.

Molecular property prediction

ML models can help determine the essential properties of an underlying molecule. These properties help to determine various aspects of a compound as well as a target that could ultimately decide how successful the drug would be. For example, in the case of compounds, you can predict their ADME properties using ML models trained on small molecule data banks. These models can then be used in the virtual screening of millions of compounds during the lead optimization phase. In the initial stages, the models can be used in hit optimization where the molecular properties determine the size, shape, and solubility of the compounds that help to determine compound quality. In the drug design workflow, these properties play the role of parameters in an optimization pipeline powered by ML models. These models can iteratively tune to come up with optimal values of these parameters so that the designed drug has the best chance of being successful. A common application of molecular property prediction is during a **quantitative structure- activity relationship (QSAR)** analysis. This analysis is typically done at the HTS stage of drug discovery and helps to achieve a good hit rate. The QSAR models help predict the biological activities of a compound based on the chemical properties derived from molecular structure. The models make use of classification and regression models to understand the category of biological activity or its value.

Molecular structure prediction

Another application of ML in drug discovery is in the prediction of structures of both large and small molecules, which is an integral part of structure-based drug design. In the case of macromolecules such as proteins, their structure defines how they function and helps you find which molecules could bind to their binding sites. In the case of small molecules, the structures help determine their kinetic and physical properties, ultimately leading to a hit. Structural properties are also critical in drug design where novel compounds are designed based on their affinity to a target structure. Traditionally, the structure of molecules has been determined using technologies such as mass spectrometry, which aims to determine the molecular mass of a compound. Another approach to determining structure, especially for more complex molecules, is crystallography, which makes use of X-rays to determine the bonding arrangements in a molecule. For macromolecules, a popular technique to determine structure is **cryo-electron microscopy (cryo-EM)**.

While these technologies show a lot of promise, ML has been increasingly used in structure prediction because of its ability to simulate these structures and study them virtually. For example, **generative models** in ML can generate possible compound structures from known properties that allow researchers to perform bioactivity analysis on a few sets of compounds. Since these compounds are already generated with known characteristics learned by the model, their chances of being successful are higher.

Language models for proteins and compounds

Language models are a type of **natural language processing** (**NLP**) technique that allows you to predict whether a given sequence of words is likely to exist in a sentence. They have wide applications in NLP and have been used successfully in applications involving virtual assistants (such as Siri and Alexa), speech-to-text services, translation services, and sentiment analysis. Language models work by generating a probability distribution of the word and picking the ones that have the most validity. The validity is decided by the rules that it learns from text corpora using a variety of techniques. It ranges from approaches such as **n-grams** for simple sentences to more complex **recurrent neural networks** (**RNNs**) that can understand context in natural language. A breakthrough discovery in language models came with the announcement of **BERT** and **GPT** models that utilize **transformer architecture**. They use an attention-based mechanism and can learn context by learning which inputs deserve more attention than others. To know more about transformer architecture, you can refer to the original paper: `https://arxiv.org/pdf/1706.03762.pdf`.

Just as language models can predict words in a sentence, the same concept has been extended by researchers in the context of proteins and chemicals. Language models can be trained on common public repositories of molecular data such as **DrugBank**, **ChEMBL**, and **ZINC** and learn the sequence of amino acids or compounds. Transformer architecture allows us to easily fine-tune the language models for specific tasks in drug discovery, such as molecule generation, and a variety of classification and regression tasks, such as property prediction and reaction prediction.

The field of drug discovery and design is ripe with innovation, and it will continue to transform the way new therapies and drugs are discovered and designed, fueled by technology and automation. It has the promise to one day find a cure for diseases such as cancer, Alzheimer's, and HIV/AIDS. Now that we have a good idea of some of the applications of ML in drug discovery and design, let's dive into the custom container options in SageMaker that allow you to use specialized cheminformatics and bioinformatics libraries for model training.

Introducing custom containers in SageMaker

In *Chapter 6* and *Chapter 7*, we went over various options for training and deploying models on Amazon SageMaker. These options allow you to cover a variety of scenarios that should address most of your ML needs. As you saw, SageMaker makes heavy use of Docker containers to train and host models. By utilizing pre-built SageMaker algorithms and framework containers, you can use SageMaker to train and deploy ML models with ease. Sometimes, however, your need may not be fully addressed by the pre-built containers. This may be because you need specific software or a dependency that cannot be directly addressed by the framework and algorithm containers in SageMaker. This is when you can use the option of bringing your own container to SageMaker. To do this, you need to adapt or create a container that can work with SageMaker. Let's now dive into the details of how to utilize this option.

Adapting your container for SageMaker training

SageMaker provides a toolkit for training to make it easy to extend pre-built SageMaker Docker images based on your requirements. You can find the toolkit at the following location: `https://github.com/aws/sagemaker-training-toolkit`.

The toolkit provides wrapper functions and environment variables to train a model using Docker containers and can be easily added to any container to make it work with SageMaker. The Docker file expects two environment variables:

- `SAGEMAKER_SUBMIT_DIRECTORY`: This is the directory where the Python script for training is located
- `SAGEMAKER_PROGRAM`: This is the program that will be executed as the entry point for the container

In addition, you can install additional libraries or dependencies into the Docker container using the `pip` command. You can then build and push the container to the ECR repository in your account. The process for training with the custom container is very similar to that of the built-in containers of SageMaker. To learn more about how to adapt your containers for training with SageMaker, refer to the following link: `https://docs.aws.amazon.com/sagemaker/latest/dg/adapt-training-container.html`.

Adapting your container for SageMaker inference

As well as a training toolkit, SageMaker provides an inference toolkit. The inference toolkit has the necessary functions to make your inference code work with SageMaker hosting. To learn more about the SageMaker inference toolkit, you can refer to the following documentation: `https://github.com/aws/sagemaker-inference-toolkit`.

The inference toolkit requires you to create an inference script and a Dockerfile that can import the inference handler, an inference service, and an entry-point script. These are three separate Python files. The job of the inference handler Python script is to provide a function to load the model, pre-process the input data, generate predictions, and process the output data. The inference service is responsible for handling the initialization of the model when the model server starts and is a method to handle all incoming inference requests to the model server. Finally, an entry-point script simply starts the model server by invoking the handler service. In the Dockerfile, we add the model handler script to the `/home/model-server/` directory in the container and also specify the entry point to be the entry-point script. You can then build and push the container to the ECR repository in your account. The process for using the custom inference container is very similar to that of the built-in containers of SageMaker. To learn more about how to adapt your containers for inference with SageMaker, refer to the following link: `https://docs.aws.amazon.com/sagemaker/latest/dg/adapt-inference-container.html`.

With this theoretical understanding of the SageMaker custom containers, let us now utilize these concepts to create a custom container to train a molecular property prediction model on SageMaker.

Building a molecular property prediction model on SageMaker

In the previous chapters, we learned how to use SageMaker training and inference using built-in containers. In this chapter, we will see how we can extend SageMaker to train with custom containers. We will be using a Dockerfile to create a training container for SageMaker using the SageMaker training toolkit. We will then utilize that container to train a molecular property prediction model and see some results of our training job in a Jupyter notebook. We will also see how to test the container locally before submitting a training job to it. This is a handy feature of SageMaker that lets you validate whether your training container is working as expected and helps you debug the errors if needed.

For the purposes of this exercise, we will use a few custom libraries. Here is a list of custom libraries that we will be using:

- **RDKit**: A collection of cheminformatics and machine-learning software written in C++ and Python: `https://github.com/rdkit/rdkit`

- **Therapeutics Data Commons (TDC)**: The first unifying framework to systematically access, evaluate, and benchmark ML methods across an entire range of therapeutics: `https://pypi.org/project/PyTDC/`

- **Pandas Flavor**: `pandas-flavor` extends `pandas`' extension API in the following ways:

  - By adding support for registering methods as well

  - By making each of these functions backward compatible with older versions of pandas: `https://pypi.org/project/pandas-flavor/`

- **Deep Purpose**: A deep learning-based molecular modeling and prediction toolkit on drug-target interaction prediction, compound property prediction, protein-protein interaction prediction, and protein function prediction (using PyTorch): `https://github.com/kexinhuang12345/DeepPurpose`

We will install these libraries on a custom training container using `pip` commands in a Dockerfile. Before we move forward, please familiarize yourselves with these libraries and their usage.

Running the Jupyter Notebook

The notebook for this exercise is saved on GitHub here: `https://github.com/PacktPublishing/Applied-Machine-Learning-for-Healthcare-and-Life-Sciences-using-AWS/blob/main/chapter-08/molecular_property_prediction.ipynb`:

1. Open the Jupyter notebook interface of the SageMaker notebook instance by clicking on the **Open Jupyter** link on the **Notebook Instances** screen.

2. Open the notebook for this exercise by navigating to `Applied-Machine-Learning-for-Healthcare-and-Life-Sciences-using-AWS/chapter-8/` and clicking on `molecular_property_prediction.ipynb`.

3. Follow the instructions in the notebook to complete the exercise.

In this exercise, we will train a molecular property prediction model on SageMaker using a custom training container. We will run the training in two modes:

- **Local mode**: In this mode, we will test our custom container by running a single model for the **Human Intestinal Absorption (HIA)** prediction model

- **SageMaker training mode**: In this mode, we will run multiple ADME models on a GPU on SageMaker

We will then download the trained models locally. At the end of the exercise, you can view the trained models in a directory called `models`.

```
! ls models/
```

```
bbb_martins_model                        cyp3a4_substrate_carbonmangels_model
bioavailability_ma_model                 cyp3a4_veith_model
caco2_wang_model                         half_life_obach_model
clearance_hepatocyte_az_model            hia_hou_model
clearance_microsome_az_model             hydrationfreeenergy_freesolv_model
cyp1a2_veith_model                       lipophilicity_astrazeneca_model
cyp2c19_veith_model                      pgp_broccatelli_model
cyp2c9_substrate_carbonmangels_model     ppbr_az_model
cyp2c9_veith_model                       solubility_aqsoldb_model
cyp2d6_substrate_carbonmangels_model     vdss_lombardo_model
cyp2d6_veith_model
```

Figure 8.1 – Trained models downloaded locally at the end of the SageMaker training job

Once you have concluded the exercise, make sure you stop or delete your SageMaker resources to avoid incurring charges, as described at the following link: `https://sagemaker-workshop.com/cleanup/sagemaker.html`.

Summary

In this chapter, we summarized the types of molecular data and how they are represented. We understood why it's important to understand molecular data and its role in the drug discovery and design process. Next, we were introduced to the complex process of drug discovery and design. We looked at a few innovations that have advanced the field over the years. We also looked at some common applications of ML in this field. For the technical portion of this chapter, we learned about the use of custom containers in SageMaker. We saw how this option allows us to run custom packages for bioinformatics and cheminformatics on SageMaker. Lastly, we built an ML model to predict the molecular properties of some compounds.

In *Chapter 9*, *Applying Machine Learning to Clinical Trials and Pharmacovigilance*, we will look at how ML can optimize the steps of clinical trials and track the adverse effects of drugs.

Further reading

- Hughes, J.P., Rees, S., Kalindjian, S.B., and Philpott, K.L. *Principles of early drug discovery* in *British Journal of Phamacology* (2011): `https://www.ncbi.nlm.nih.gov/pmc/articles/PMC3058157/`

- *Small Molecule*: `https://www.sciencedirect.com/topics/biochemistry-genetics-and-molecular-biology/small-molecule`

- Yasmine S. Al-Hamdani, Péter R. Nagy, Dennis Barton, Mihály Kállay, Jan Gerit Brandenburg, Alexandre Tkatchenko. *Interactions between Large Molecules: Puzzle for Reference Quantum-Mechanical Methods*: `https://arxiv.org/abs/2009.08927`

- Dara S, Dhamercherla S, Jadav SS, Babu CM, Ahsan MJ. *Machine Learning in Drug Discovery: A Review*. Artif Intell Rev. 2022;55(3):1947–1999. doi: 10.1007/s10462-021-10058-4. Epub 2021 Aug 11. PMID: 34393317; PMCID: PMC8356896. `https://pubmed.ncbi.nlm.nih.gov/34393317/`

9

Applying Machine Learning to Clinical Trials and Pharmacovigilance

Clinical trials or clinical research is a crucial phase in the process of taking a drug or therapy to market. Before the clinical trial phase, the drug is tested in labs and on animals only. At the end of the pre-clinical research phase, highly promising candidates are identified, and they move to the clinical trial phase. This is the first time the drug or therapy is administered to humans. The clinical trial phase provides evidence that the drug or therapy is safe enough to be administered in humans and also has the desired effects (**efficacy**). The process can take anywhere between 10 and 15 years and follows strict guidelines and protocols. It is also a huge investment from drug manufacturers who spend billions of dollars on execution and support for clinical trials. Clinical trials can also be funded by government agencies and **academic medical centers (AMCs)** for research purposes, such as observing the effects of certain drugs on certain cohorts of patients. They are known as the **sponsors** of the clinical trial.

As you might have guessed, clinical trials are heavily controlled and regulated by the **Food and Drug Administration (FDA)** and the **National Institutes of Health (NIH)**, who have strict rules for such studies. Everything from the place where the trials are conducted to the participants of the trial are pre-defined and carefully selected. The trials are supported by doctors, nurses, researchers, and a variety of other medical professionals who have specific roles. There are strict measures in place to maintain the confidentiality of participants during the trial ensuring only authorized personnel have access. The data gathered during clinical trials, especially the adverse events, have strict reporting guidelines and need to be submitted to the FDA and NIH in a timely manner.

Because of their highly controlled conditions, successful clinical trials are not a guarantee that the drug will perform as intended throughout its life cycle when administered to a larger number of patients. In fact, drugs need to be monitored throughout their life cycle for **adverse effects**. The process of detection, understanding, and prevention of drug-related problems, and then reporting this data to the authorities, is known as **pharmacovigilance (PV)**.

In this chapter, we will dive into the clinical trial process. We will look at the important steps that a drug needs to go through before it is released for general use. Additionally, we will gain an understanding of how the agencies ensure regulations are followed and how they keep the public safe from adverse effects:

- Understanding the clinical trial workflow
- Introducing PV
- Applying ML to clinical trials and PV
- Introducing SageMaker Pipelines and Model Registry
- Building an adverse event clustering model pipeline on SageMaker

Technical requirements

The following are the technical requirements that you need to complete before building the example implementation at the end of this chapter:

1. Complete the steps to set up the prerequisites for Amazon SageMaker as described here: `https://docs.aws.amazon.com/sagemaker/latest/dg/gs-set-up.html`.

2. Create a SageMaker notebook instance by following the steps in the following guide: `https://docs.aws.amazon.com/sagemaker/latest/dg/howitworks-create-ws.html`.

3. Create an S3 bucket, as described in *Chapter 4*, in the *Building a smart medical transcription application on AWS* section under *Create an S3 bucket*. If you already have an S3 bucket, you can use that instead of creating a new bucket.

4. Open the Jupyter notebook interface of the SageMaker notebook instance by clicking on the **Open Jupyter** link from the **Notebook Instances** screen.

5. In the top-right corner of the Jupyter notebook, click on **New** and then **Terminal**.

6. Type in the following commands on the terminal screen:

```
$cd SageMaker
$git clone https://github.com/PacktPublishing/Applied-
Machine-Learning-for-Healthcare-and-Life-Sciences-using-
AWS.git
```

You should now see a folder named `Applied-Machine-Learning-for-Healthcare-and-Life-Sciences-using-AWS`.

> **Note**
>
> If you have already cloned the repository in a previous exercise, you should already have this folder. You do not need to do *step 6* again.

7. Onboard to SageMaker Studio Domain using the quick setup described at `https://docs.aws.amazon.com/sagemaker/latest/dg/onboard-quick-start.html`.

> **Note**
>
> If you have already onboarded a SageMaker Studio Domain from a previous exercise, you do not need to perform *step 7* again.

8. Familiarize yourself with the SageMaker Studio UI components: `https://docs.aws.amazon.com/sagemaker/latest/dg/studio-ui.html`.

Once you have completed these steps, you should be all set to execute the steps from the example implementation in the last section of this chapter.

Understanding the clinical trial workflow

Clinical trials are needed for any type of new medical intervention on humans. This includes drugs, therapies, certain types of medical devices, and procedures. The trials follow a plan or a **protocol** that is defined by the **investigator** in a **protocol document**. Depending on the goals of the trial, a clinical trial might aim to compare the **safety** and **efficacy** of a product (such as a drug or a device) or a procedure (such as a new therapy) by comparing it against a standard product or a procedure. In some cases, participants are given a drug with no active ingredient, also known as a **placebo**, to compare them against participants who receive the new drug. This helps determine whether the drug is effective and is working as expected. For example, a new drug being studied to lower blood sugar is given to some participants who have high blood sugar. This is compared to another set of participants who are given a placebo. The observations from the two sets of participants are monitored to see whether the participants who were given the new drug have normal sugar levels or not. In addition, the participants are monitored for any reactions or side effects as a result of taking this new drug. These are known as **adverse events** and need to be carefully monitored throughout the life cycle of the trial. Adverse events could include minor issues such as nausea, rashes, or headache. There are serious adverse events that might lead to hospitalization or might cause disability, permanent damage, or even death. Principal investigators of the trials ensure they take necessary steps and follow all regulatory aspects and ensure there are no serious adverse events in the trial.

Participants for a trial are carefully chosen based on the goals for the trial, as defined in the protocol. Participants must meet the eligibility criteria, also known as the **inclusion criteria** for a trial. Additionally, the protocol has details on **exclusion criteria** for the trial, which may disqualify a participant from the trial. Inclusion and exclusion criteria include demographic information such as age and gender. It might also include specific clinical criteria such as the stage of a certain disease or medical history.

Participation in clinical trials is mostly voluntary, and the protocol defines the number of participants required in the trial. Before becoming a participant in a clinical trial, a person must sign a document that acknowledges that they know the details of the trial and clearly understand their risks and potential benefits. This is known as **informed consent**. The participants sign the **informed consent document** and are free to drop from the trial at any time, even before the trial is complete. In general, the informed consent process is designed to protect the participants in the clinical trial. There are multiple review boards and agencies that review the study protocol, the informed consent document, and the overall study design to ensure the safety of the clinical trial participants. One such body is the **institutional review board (IRB)**. They are responsible for reviewing all federally supported clinical studies in the US and consist of physicians, researchers, and members of the community.

Clinical trials help progress the medical field and help bring new drugs and therapies to market that could have profound benefits for future patients. It may have some risks, but the regulators review those carefully and ensure the risks are fairly reasonable when compared to the benefits.

The clinical trial workflow involves multiple steps that are carried out in phases. Now, let's look at the various steps in more detail by taking a clinical trial for a new drug as an example:

- **Study design**: During the study design phase, the researchers determine the goals of the trials and the specific questions they are looking to answer. This includes the inclusion and exclusion criteria of the trial, the number of people who can participate, the length and schedule of the trial, and the dosage and frequency of the drugs. It also defines what data will be collected, how often it will be collected, and the rules of analysis and interpretation of the data. This is all available in the trial protocol document that is followed by the clinical trial.

- **Phase 1**: This is the phase when researchers test the drug on a few participants, usually less than 100. This phase determines how the new drug interacts with the body and carefully monitors the participants for side effects. It also determines the correct dosage of the drug and gathers initial data about how effective the drug is in a small group of participants.

- **Phase 2**: In phase 2, there are a few hundred patients recruited who have specific clinical conditions that the drug is expected to treat. The main purpose of phase 2 of the study is to determine the safety of the drug. The inputs from phase 2 help in the larger phase 3 study.

- **Phase 3**: This phase involves hundreds or even thousands of participants. This phase tries to mimic a real-world scenario of administering the drug to large portions of a population and determines whether the drug is beneficial. Because this is a much larger study with many more participants, this phase is able to detect adverse effects that might not have been evident in smaller groups participating in phases 1 and 2. The phase lasts several years and also determines the long-term effect of the drug.

- **Phase 4**: Although clinical trials are only conducted in these three phases for authorization by the FDA, sometimes a fourth phase of the clinical trial is also referenced. This refers to the post-market launch of the drug and how it performs on thousands of patients who are prescribed the new drug by doctors.

Before a clinical trial can begin in the US, the sponsors submit an application to the FDA to seek approval to begin the trial. This is known as the **Investigational New Drug (IND)** application process. The manufacturers include all information about the drug, the protocols, any relevant prior data, and details about the investigator for the trial. The FDA offers help to sponsors related to the submission of the IND application and answers any questions the sponsors might have in relation to the application. Once the application is submitted, the FDA has a certain amount of time (usually 30 days) to respond with its decision.

The US national library of medicine maintains a list of clinical trials at `https://clinicaltrials.gov/`. This is a resource where you can search through a variety of information about specific clinical studies such as their location, their type, their status, and details of the protocols.

While randomized controlled trials provide the necessary information to regulators about the safety and efficacy of the drug in a population with strict inclusion and exclusion criteria, sometimes, it is desirable to study the drug in non-controlled settings in a more diverse patient population. This is where **real-world evidence (RWE)**-based studies come in. Next, let's understand RWE in more detail.

Understanding RWE studies

RWE studies are studies designed to understand the long-term effects of a drug on the wider population in uncontrolled settings, mimicking the use of the drug in the real world. RWE studies are done by collecting **real-world data (RWD)**, which is not typically collected in a controlled clinical trial. RWD is passively collected from different sources such as **electronic medical records (EMRs)**, claims and billing systems, prescriptions, and wearable devices. The data is then analyzed over a period of time to help regulators and healthcare providers understand how effective and safe the drug is in patients in their real-world treatment plans. It helps generalize the drug's effects in a more realistic population of patients.

RWE studies have a wide application both in the pre- and post-launch stages of a new drug. The outcomes from these studies are beneficial for healthcare providers, payers, and regulators. In the pre-launch phase, it helps decide the specifics of the patient population to be included in the trial and helps payers and providers determine the correct treatment plan. Post-launch, it can help determine the correct pricing and coverage for the drug by payers. I can also help physicians determine the right messaging and value proposition for the drug.

Because of its wide scope of data collection and analysis, RWE studies are faced with the challenge of data integration and analysis. In a few cases, the data might be incomplete and might have inconsistencies such as duplicate records. Deriving meaningful information from RWD that is useful for policymakers and regulators involves robust data analysis and processing techniques. Technologies such as **big data analytics**, **natural language processing**, **speech to text**, and even **computer vision** play a big role in the analysis and interpretation of RWE.

Now that we have an understanding of the clinical trial process, let us look at PV in more detail.

Introducing PV

Before a drug is launched for use, it is tested in small populations under controlled settings, which prevents us from understanding how the drug would behave in the long term on large groups of people. PV refers to the activities and science related to the detection, assessment, and prevention of adverse effects related to medicines. It is designed to improve patient safety and confidence around the use of the drug. During clinical research, PV is conducted by monitoring the patients closely in controlled environments. Post-launch, PV is carried out via observational studies and **post-market surveillance** (**PMS**). While the data collected during a clinical trial is better in quality, it is quite limited. Hence, the post-market PV attempts to draw conclusions from a wider real-world dataset in uncontrolled settings.

The central element of post-market PV revolves around the concept of monitoring drug usage in the real world for extended periods of time. During the monitoring process, the data around adverse effects and the severity of those adverse effects are captured. Ultimately, the data is made available to healthcare professionals and patients to generate awareness about the drug's risks. It may also need to be reported to regulatory agencies. For example, the safety issues related to FDA-regulated products are reported to the FDA using an **individual case safety report** (**ICSR**). PV data also helps policymakers and regulators understand trends and approve drug usage in new markets. It helps researchers carry out new R&D activities that improve the safety and efficacy of the drug and even develop new ones.

Now, let us understand how post-market PV is conducted by diving into the different stages of the PV process. There are two broad areas that the PV process can be divided into.

Signal generation

Signal generation is the process of identification of a new **adverse drug reaction** (**ADR**). One major source of signal generation is **spontaneous reporting**. Spontaneous reporting is the voluntary recording and sharing of clinical observations about ADR due to drug usage. This is done by physicians, pharmacists, and even pharmaceutical companies. The reports are investigated by PV staff and are maintained in a database. Since it is a voluntary process, the culture of reporting ADRs via spontaneous reporting varies greatly. It is also difficult to ascertain whether the ADR is due to the reaction from the drug or some other reason in the patient's care plan. This is also known as **signal versus noise** and is a key challenge in signal generation. It does have its benefits because data about the ADR is immediately available following the drug launch. Another approach to signal detection is via case reports published in journals that alert physicians about possible hazards of a drug. This method is more rigorous but has limitations, as not all ADRs can be published in journals. In some cases, specialized cohort studies might be set up to generate signals about a new possible ADR. However, because these studies deviate from the real world and have a limited size, it does not guarantee successful signal generation. Organizations might also carry out large randomized trials to study safety concerns about drugs. Once a signal has been generated, the process moves toward the hypothesis testing phase.

Hypothesis testing

Once a concerning report about an ADR associated with a drug arises in the signal generation process, the next step is to prove that the ADR is caused by the drug and not any other factor. This is known as hypothesis testing. The key here is to assess causality between the clinical conditions in individual cases and the taking of a certain drug. The data from spontaneous reporting can be used to assess the hypothesis. For example, if the spontaneous reporting indicates a large number of similar ADRs associated with the drug, it can be further investigated. This is done via epidemiological studies that are observational in nature and improve our understanding of the drug's safety. Once an ADR has been established, drug manufacturers may choose to include this information in the prescribing material of the drug. As you can imagine, including information about a drug having adverse reactions in the prescribing material of the drug can have huge implications. It might deprive patients of receiving the treatments that they need. It can also result in a lot of negative attention regarding the reliability of the drug if the reported events are of a serious nature. Therefore, it is extremely critical for the hypothesis to be fully tested and established before being associated with the drug.

The PV practice continues to be modernized with the use of technology in the detection and reporting of adverse reactions. However, the key ask remains the same: *how can we make drugs safer for patients and reduce the risk of serious adverse reactions?* Now, let us look at how ML can be utilized in clinical trials and PV to help reduce the risk to patients and also make the workflow more efficient.

Applying ML to clinical trials and PV

The clinical trial process is mostly sequential and generates large volumes of data in different modalities. These datasets need to be analyzed and processed for information retrieval. The information embedded in these data assets is critical to the success of the overall trial. The information from a previous step might inform how future steps need to be carried out. Moreover, regulatory requirements emphasize strict timelines and governance regarding how you process this information. ML can help automate the repeatable steps in a clinical trial process and help make the trial more efficient. Additionally, it can help with information discovery to better inform regulators and policymakers about future trials. Now, let us look at some common ways in which ML can be applied to clinical trials and PV.

Literature search and protocol design

During the study design phase, scientists need to search through a variety of literature from previous clinical trials to find similar compounds. The information is embedded in multiple documents in a variety of formats. It is also distributed in different sources such as remote file systems, websites, or content portals. ML-based information search and discovery can help crawl for information embedded within these documents and make them accessible to researchers via simple search queries in natural language. One common technique utilized in these search algorithms is **graph neural networks**. They allow you to train deep learning models on a graph representation of the underlying data.

Information retrieved from these searches can help you to design accurate clinical studies for new trials that take into account the findings and learnings from previous trials. AWS services such as **Amazon Kendra** allow you to easily create a graph-based search engine customized for your documents. To learn more about Amazon Kendra, refer to the following link: `https://aws.amazon.com/kendra/`.

Trial participant recruitment

A key step in the clinical trial is the recruitment of the right trial participants. As described earlier, trial participants are carefully selected based on the inclusion and exclusion criteria defined in the protocol document. Another key aspect to keep in mind while selecting the trial participants is their chances of dropping out from the trial. Participation is voluntary and participants are free to drop out at any time, even in the middle of the trial. Mid-trial dropouts can jeopardize the whole trial and delay the process considerably. Therefore, it is important to take multiple aspects into account, beyond just the inclusion and exclusion criteria, before selecting the participants. They could include the participants' ability to travel to the trial site based on the distance of their residence and the site, the dependents they need to care for while they are participating in the trial, their ability to adhere to the schedule of the trial, and more. These aspects are captured using forms and surveys that volunteers need to fill out. The manual analysis of responses from these forms is time-consuming. Using ML can help automate the analysis of clinical trial recruitment forms and make the process more efficient. For example, **named entity recognition** (**NER**) models can recognize specific entities in the clinical trial form that are crucial for patient selection. **Classification models** can then classify possible participants or a **regression model** can compute a risk score for them to denote how likely they are to drop out mid-trial.

Adverse event detection and reporting

Safety is the highest priority in any clinical trial. Hence, there are strict regulations in place to report any safety issues such as adverse events to regulatory agencies in a timely manner. Adverse events in clinical trials could be reported in a variety of ways. For example, an ICSR might be used to report specific adverse events for individuals participating in the trial to the FDA. The information about adverse events comes from multiple sources. It might be self-reported by the participants or by the clinicians who are monitoring the patients. It might be reported by mediums such as emails, paper forms, or voice. As a result, it is becoming difficult to gather information about the safety of drugs in a clinical trial and report it in a timely manner to the regulatory agencies. ML models can be trained on safety data to identify possible adverse events from different sources of data. These models can then be applied to data generated in a clinical trial to detect adverse events. It can also monitor for the occurrence of adverse events in real time for participants in a remote trial. It might then be sent to investigators to manually verify and send to regulatory agencies. The process creates a more efficient way of detecting and reporting adverse events.

Real-world data analysis

During the post-market surveillance phase of the drug, real-world data needs to be continuously monitored. It is a crucial phase for the drug as it is being administered by doctors to real patients in uncontrolled environments for the first time. Unwanted outcomes in this phase could result in recalls for the drug, which can prove to be extremely costly for drug manufacturers. As described earlier, RWD is gathered from multiple sources such as EHR systems and payment/claims systems. RWD can also be collected from social media accounts and drug review websites. All this makes data analysis and the capturing of meaningful outcomes extremely difficult from RWD. ML has been successfully used on RWD for PV purposes. Classification models can help classify patients at risk of unwanted clinical outcomes as a result of taking the drug. These models can be built on multi-modal datasets such as voice, images, and clinical notes, which makes these models adapt to real-world scenarios. Ultimately, the predictive models help us to identify events before they happen, giving clinicians enough time to prevent them or make their impact as low as possible to help keep the patients safe.

As you can see, ML has an impact on all stages of the clinical trial and PV process. It helps design smarter trials, execute them more efficiently, and reduce or eliminate adverse events. Now, let us look at a feature in SageMaker that helps us create model pipelines. We will then use this to create an adverse events clustering model pipeline in the last section of the chapter.

Introducing SageMaker Pipelines and Model Registry

In previous chapters of this book, you were introduced to different options in SageMaker to process data, extract features, train models, and deploy models. These options provide you with the flexibility to pick the components of SageMaker that work best for your use case and stitch them together as a workflow. In most cases, these workflows are repeatable and need to be executed in different environments. Hence, you need to maintain them using an external orchestrating tool that helps you design the workflow and maintain it for repeated runs. This is where SageMaker Pipelines comes in.

SageMaker Pipelines is a model-building pipeline that allows you to create a visual **directed acyclic graph** (**DAG**) for the various steps of your model-building process and manage it as a repeatable workflow. The DAG is exported in **JSON** format and provides details about relationships between each step in the pipeline. You can pass the output of one step in your DAG as an input to another step, creating a sequential data dependency between the steps. This is then visually represented to make it easy for you to understand the flow of the pipeline and its various dependencies.

To make it easy for you to define the pipeline, SageMaker provides a Pipelines SDK. You can see more details about the Pipelines SDK at `https://sagemaker.readthedocs.io/en/stable/workflows/pipelines/sagemaker.workflow.pipelines.html`.

Let us now understand the construct of a basic pipeline and the functionalities it can support.

Defining pipeline and steps

A SageMaker pipeline consists of multiple steps and parameters. The order of execution of the steps is determined by the data dependencies between the steps and is automatically inferred.

Here is an example of a pipeline definition using the SageMaker Pipelines SDK in Python:

```
from sagemaker.workflow.pipeline import Pipeline
pipeline_name = f"mypipeline"
  pipeline = Pipeline(
      name=pipeline_name,
      parameters=[
          processing_instance_type,
          processing_instance_count,
          training_instance_type,
          model_approval_status,
          input_data,
          batch_data,
      ],
steps=[step_process, step_train, step_eval, step_cond],
  )
```

As you can see from the preceding code definition, the pipeline accepts a name that must be unique for each region. It also accepts steps in the pipeline such as processing and training. Lastly, the pipeline has a set of parameters such as input data and an instance count.

In addition to these parameters, you can also pass variables in your pipeline steps using pipeline parameters. Here is an example of how to do that:

```
from sagemaker.workflow.parameters import (
    ParameterInteger,
    ParameterString,
    ParameterFloat,
    ParameterBoolean
)

processing_instance_count = ParameterInteger(
    name="ProcessingInstanceCount",
    default_value=1
)
```

```
pipeline = Pipeline(
    name=pipeline_name,
    parameters=[
        processing_instance_count
    ],
    steps=[step_process]
)
```

As you can see from the preceding code, we defined a parameter called ProcessingInstanceCount and passed it as input when creating the pipeline.

A pipeline supports multiple types of steps that are common in a ML workflow. Some of the common steps are **processing**, **training**, **tuning**, and **CreateModel**. For a full list of steps and example code showing how to use them, take a look at https://docs.aws.amazon.com/sagemaker/latest/dg/build-and-manage-steps.html#build-and-manage-steps-types.

Caching pipelines

Sometimes, you need to run the same step with the same parameters multiple times when running a pipeline. SageMaker pipeline caching enables you to reuse the results from a previous execution, so you do not need to run the repeatable steps again. It checks for the previous execution with the same invocation parameters and creates a cache hit. Then, it utilizes the cache hit for subsequent executions instead of running the step again. This feature is available for a subset of steps supported by SageMaker Pipelines. To enable step caching, you must define a CacheConfig property in the following manner:

```
{
    "CacheConfig": {
        "Enabled": false,
        "ExpireAfter": "<time>"
    }
}
```

Enabled is a Boolean and is false by default. The ExpireAfter string accepts any ISO 8601 duration string. Here is an example of how to enable caching for a training step:

```
from sagemaker.workflow.pipeline_context import PipelineSession
from sagemaker.workflow.steps import CacheConfig

cache_config = CacheConfig(enable_caching=True, expire_
after="PT1H")
estimator = Estimator(..., sagemaker_session=PipelineSession())
```

```
step_train = TrainingStep(
    name="TrainAbaloneModel",
    step_args=estimator.fit(inputs=inputs),
    cache_config=cache_config
)
```

The preceding code block enables caching for one hour on the training step.

There are other useful features within SageMaker Pipelines that allow you to create complex workflows and pass parameters between steps. Its flexible nature allows you to integrate multiple custom steps together. For a full review of the SageMaker Pipelines features, you can refer to the SageMaker developer guide at https://docs.aws.amazon.com/sagemaker/latest/dg/pipelines.html.

Often, it is the case that a pipeline might need to be run multiple times to produce different versions of the same model. Some of these models might not be suitable for deployment to production as their accuracies are not suitable. It is essential to maintain all versions of the model so that the ones ready for deployment can be approved and pushed to production. The **SageMaker Model Registry** allows you to do this. Let us look at the SageMaker Model Registry in more detail.

Introducing the SageMaker Model Registry

The SageMaker Model Registry supports model groups, which is a way to group together multiple versions of a model for the same use case. The models are maintained in a **model package** with an associated version number. The model package has the trained model artifacts as well as the inference code that is used to generate inference from the model. A model needs to be registered in a model group before it can be deployed. The registered model is part of a versioned model package that is maintained within the model group. You can view details of any model package using the `list_model_packages` method that is available in the **SageMaker Python SDK**. Additionally, you can delete a particular version if it doesn't apply to the model group.

Another important feature of the SageMaker Model Registry is the ability to update the status of the model. There are three statuses that the model can be: **Approved**, **Rejected**, or **PendingManualApproval**. By default, the models are in **PendingManualApproval**. Here are the possible transitions that the model can have and their associated effects:

- **PendingManualApproval to Approved** – This initiates CI/CD deployment of the approved model version

- **PendingManualApproval to Rejected** – No action

- **Rejected to Approved** – This initiates CI/CD deployment of the approved model version

- **Approved to Rejected** – This initiates CI/CD to deploy the latest model version with an **Approved** status

Once the model has been approved, it is ready to be deployed as an endpoint for inference. These deployments are also allowed across accounts. This is handy when you want to move a model from a staging account (such as dev or test) to a production account. To learn more about the SageMaker Model Registry, please refer to the developer guide at `https://docs.aws.amazon.com/sagemaker/latest/dg/model-registry.html`.

Now that we understand the basics of SageMaker Pipelines and the Model Registry, let us use these concepts to build a pipeline for clustering adverse events associated with drugs.

Building an adverse event clustering model pipeline on SageMaker

Now let us build a pipeline to train an adverse event clustering model. The purpose of this pipeline is to cluster adverse events detected in drug reviews using an unsupervised clustering model. This can help investigators group drugs with certain reported clinical conditions together and facilitates investigations related to adverse events. We will read some raw drug review data and extract top clinical conditions from that data. Let us now look at the details of the workflow. Here is a diagram that explains the steps of the solution:

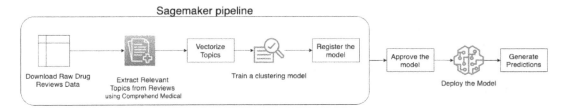

Figure 9.1 – The pipeline workflow

As shown in the preceding diagram, we use **Amazon Comprehend Medical** to extract clinical conditions from the raw drug reviews. These clinical conditions are reported by the end users as adverse events while taking the drug. We take the top five clinical conditions as relevant topics on which we would like to cluster. Clustering is an unsupervised learning technique that allows us to derive groupings from unlabeled data. For the purposes of this solution, we randomly choose 100 drugs from the raw dataset to limit the execution time for the processing step.

Next, we vectorize the extracted topics for each drug using a **term frequency–inverse document frequency** (**tf–idf**) vectorizer. This generates numerical representations of the topics, based on the frequency with which they occur in the corpus. For this step, we create a custom Docker container and run it on SageMaker Processing. To learn more about tf–idf, you can refer to `https://en.wikipedia.org/wiki/Tf%E2%80%93idf`.

Lastly, we create two clusters from these vectors to separate the topics into two groups. We deploy the model as a SageMaker endpoint and generate predictions for each drug in our sample.

The raw data processing, training, and registration of the trained model are part of a SageMaker pipeline. We will run the pipeline as a single execution that will read 100 drug reviews from the raw data, extract topics from each review, vectorize the topics, and train and register the model on the SageMaker Model Registry. Then, we will approve the model for deployment and deploy the model as a real-time endpoint to generate our cluster predictions for each drug.

The raw dataset for this application is available for download from the UCI ML repository at `https://archive.ics.uci.edu/ml/datasets/Drug+Review+Dataset+%28Drugs.com%29`.

You do not need to download this dataset manually. This is done by the processing script when we run it in the pipeline. So, let's get started!

Running the Jupyter notebooks

The notebooks for this exercise are saved on GitHub at `https://github.com/PacktPublishing/Applied-Machine-Learning-for-Healthcare-and-Life-Sciences-using-AWS/tree/main/chapter-09`.

As you can see, we have two notebooks for this exercise. The first one creates the pipeline, and the second one deploys the model and generates predictions. We will also use the SageMaker Model Registry to approve the model before running the second notebook:

1. Open the Jupyter notebook interface of the SageMaker notebook instance by clicking on **Open Jupyter** from the **Notebook Instances** screen.

2. Open the first notebook for this exercise by navigating to `Applied-Machine-Learning-for-Healthcare-and-Life-Sciences-using-AWS/chapter-9/` and clicking on `adverse-reaction-pipeline.ipynb`.

3. Follow the instructions in the notebook to complete the exercise.

Once you have completed the steps of the first notebook, you have started a pipeline execution for our model pipeline. Let us use the SageMaker Studio interface to look at the details of our pipeline.

Reviewing the pipeline and model

Let's take a closer look at our pipeline:

1. Launch SageMaker Studio from the AWS console.

2. In the left-hand navigation pane, click on the **SageMaker resources** icon, as shown in the following screenshot:

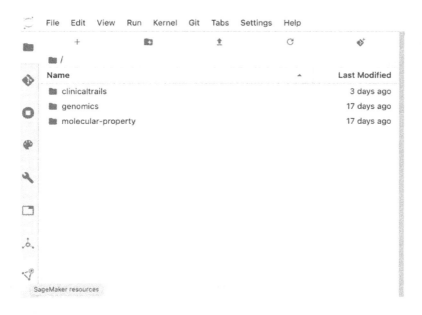

Figure 9.2 – The SageMaker resources icon in SageMaker Studio

3. Select **Pipelines** from the drop-down menu. You should see a pipeline named **adverse-drug-reaction**, as shown in the following screenshot:

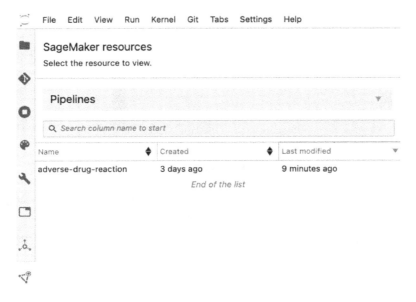

Figure 9.3 – The SageMaker pipeline screen

4. Double-click on the **adverse-drug-reaction** pipeline. This will open up all the executions of the pipeline on the right-hand window.

5. Click on the **Graph** tab at the top. This will show you a visual representation of your pipeline, as shown in *Figure 9.4*:

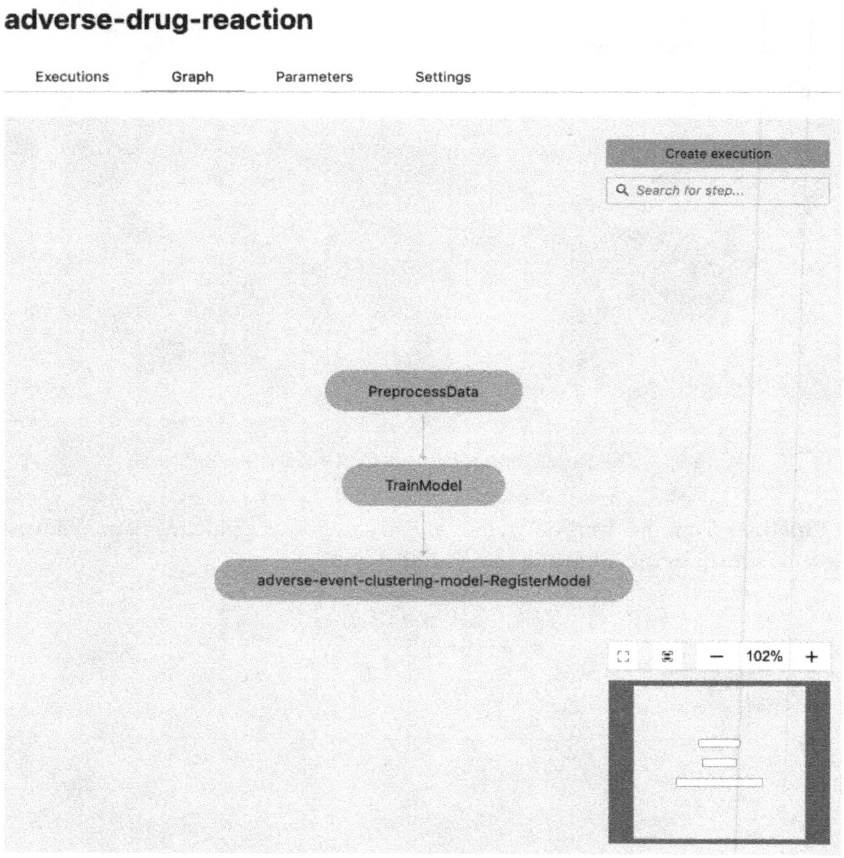

adverse-drug-reaction

Figure 9.4 – The SageMaker pipeline graph

As you can see, our pipeline consists of three sequential steps to preprocess data, train the model, and register the model in the Model Registry.

6. Go back to the **Executions** tab at the top. Double-click on the latest execution. This will open up the details of the execution.

7. You might still have your pipeline in execution mode. Wait for all three steps to complete. Once the steps are complete, they should turn green. Examine the steps in the execution by clicking on them, as shown in the following screenshot:

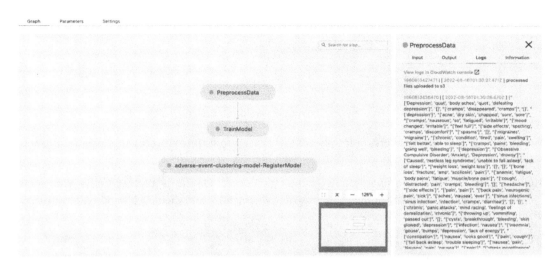

Figure 9.5 – The PreprocessData step details

As you can see in *Figure 9.5*, you can look at the details of each step in our pipeline by clicking on them. For example, here I am looking at the logs from my preprocessing step.

Now let us look at the model that was registered by our pipeline in the SageMaker Model Registry. We will also approve the model for deployment after we examine it.

8. Click on **SageMaker resources** in the left-hand navigation pane of SageMaker Studio. Select **Model Registry** from the drop-down menu. Double-click on the model group name: **adverse-event-clustering**. This will open up the models in this package, as shown in the following screenshot:

adverse-event-clustering

Version	Stage	Status	Short description	Modified by	Last modified	Actions
9	None	Pending				···
8	None	Approved			36 minutes ago	···
7	None	Approved			52 minutes ago	···
6	None	Approved			2 hours ago	···
5	None	Approved			2 hours ago	···
4	None	Approved			3 hours ago	···
3	None	Approved		ujjwalr	3 hours ago	···
2	None	Approved			2 days ago	···
1	None	Approved			2 days ago	···

Figure 9.6 – Model versions

9. The latest version of the model will show the status as **Pending**. Double-click on it. This will open the model details screen:

Figure 9.7 – The model version status

In my case, the model version is **9**, but if this is your first time running this pipeline, your model version will be **1**.

10. Click on the **Update status** button in the top right corner. On the next screen, select **Approve** from the **Status** drop-down menu, as shown in *Figure 9.10*, and then click on **Update status**:

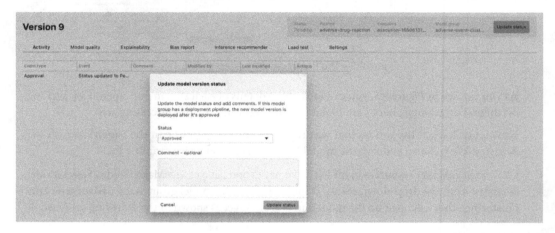

Figure 9.8 – Update model version status

We have just approved our model to be deployed for inference. You should now see that the status of the model is **Approved** in the model package screen that we accessed in *step 11*.

Next, we will deploy the approved model, run some predictions on it, and examine our results via a scatter plot.

Deploying model and running inference

The second notebook deploys the model and generates predictions. Before you perform the following steps, make sure the model in the registry is in the **Approved** state:

1. Open the Jupyter notebook interface of the SageMaker notebook instance by clicking on **Open Jupyter** from the **Notebook Instances** screen.

2. Open the first notebook for this exercise by navigating to `Applied-Machine-Learning-for-Healthcare-and-Life-Sciences-using-AWS/chapter-9/` and clicking on `adverse-reaction-inference.ipynb`.

3. Follow the instructions in the notebook to complete the exercise.

At the end of this notebook, you should see a scatter plot showing the clustering model inference output for the adverse reaction inputs. The plot should look similar to the following:

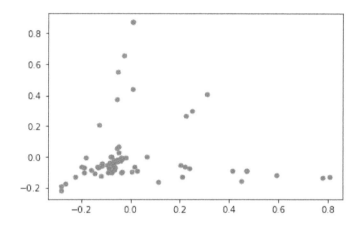

Figure 9.9 – Sample output of the clustering model

This concludes our exercise. Please make sure you stop or delete your SageMaker resources to avoid incurring charges, as described at `https://sagemaker-workshop.com/cleanup/sagemaker.html`.

Summary

In this chapter, we went into the details of how a new drug is tested for safety and efficacy before it can be launched in the market. We understood the various phases in the clinical trial workflow and looked at how regulatory agencies make policies to ensure the safety of patients and trial participants. We understood the importance of PV in the overall monitoring of the drug and looked into the details of real-world data. Additionally, we learned about how ML can optimize the clinical trial workflow and make it safer and more efficient. Finally, we learned about the new features of SageMaker called SageMaker Pipelines and Model Registry, which can aid in these processes. We also built a sample workflow to cluster adverse event data about drugs.

In *Chapter 10, Utilizing Machine Learning in the Pharmaceutical Supply Chain*, we will look at how pharma manufacturers are utilizing ML to maximize the return on multi-year investments and launching a new drug on the market.

10

Utilizing Machine Learning in the Pharmaceutical Supply Chain

The pharmaceutical supply chain refers to the process and components involved in the manufacturing and delivery of drugs that patients need promptly. It is a complicated process that involves synchronization between multiple entities and systems. There are multiple stakeholders along the way, each with its own success criteria that define how the drug gets manufactured, distributed, and prescribed. The prescription drugs are sent directly to pharmacies that distribute them to patients in accordance with their prescriptions. Nonprescription drugs or **over-the-counter** (**OTC**) drugs are sold directly by retailers to consumers. In both cases, demand forecasting for the drug has a big influence on its manufacturing and distribution. Knowing how much of a particular drug would be needed at a particular pharmacy or store helps optimize manufacturing and distribution pipelines to ensure they can keep up with demand.

To ensure pharmaceutical manufacturers have the best returns on their investments and improve profitability, it is important for them to reinforce demand generation for their brand of drugs in an increasingly competitive market. One of the ways they do this is by investing heavily in sales and marketing for the drug. They allocate large budgets for marketing their drugs in the media and also hire large teams of pharmaceutical sales representatives who are compensated with commission for selling their drugs to physicians who can prescribe them and retailers who can sell them over the counter. Another important factor to consider is the price point of the drug. This has to be carefully balanced to take into account the investments made by the pharmaceutical companies in the research and manufacturing of the drug and also the profit margins they expect. There has been increasing scrutiny by policymakers on drug prices, with the increasing price of prescription drugs controlled largely by for-profit pharmaceutical companies.

ML plays a big role in making sure that drugs are manufactured in the right volume while reducing waste and are distributed based on demand, and that the right physicians and patients are targeted to ensure maximum success. It is also essential to track market performance based on which pharmaceutical companies can take timely corrective action in case things do not go as expected. In this chapter, we will understand the pharmaceutical supply chain and get into the details of how it functions. We will look at the important ways in which ML can optimize the pharmaceutical supply chain and ensure it's run in the most optimal manner.

This chapter will cover the following topics:

- Understanding the pharmaceutical supply chain landscape
- Applying ML to the pharmaceutical supply chain and sales
- Introducing Amazon Forecast
- Building a pharmaceutical sales forecasting model using Amazon Forecast

Technical requirements

You need to complete the following technical requirements before building the example implementation at the end of this chapter:

1. Complete the steps to set up the prerequisites for Amazon SageMaker, as described here: `https://docs.aws.amazon.com/sagemaker/latest/dg/gs-set-up.html`.

2. Create a **Simple Storage Service** (**S3**) bucket, as described in *Chapter 4*, in the *Building a smart medical transcription application on AWS* section, under the *Creating an S3 bucket* heading. If you already have an S3 bucket, you can use that instead of creating a new bucket.

3. Onboard to SageMaker Studio Domain using the quick setup, as described here: `https://docs.aws.amazon.com/sagemaker/latest/dg/onboard-quick-start.html`.

> **Note**
> If you have already onboarded a SageMaker Studio domain from a previous exercise, you do not need to perform *step 3* again.

4. Once you are in the SageMaker Studio interface, click on **File** | **New** | **Terminal**.

5. Once in the terminal, type the following command:

```
git clone https://github.com/PacktPublishing/Applied-
Machine-Learning-for-Healthcare-and-Life-Sciences-using-
AWS.git
```

You should now see a folder named `Applied-Machine-Learning-for-Healthcare-and-Life-Sciences-using-AWS`.

> **Note**
> If you have already cloned the repository in a previous exercise, you should already have this folder. You do not need to do *step 5* again.

6. Familiarize yourself with the SageMaker Studio UI components: `https://docs.aws.amazon.com/sagemaker/latest/dg/studio-ui.html`.

Once you have completed these steps, you should be all set to execute the steps in the example implementation in the last section of this chapter.

Understanding the pharmaceutical supply chain landscape

The overall pharmaceutical supply chain landscape is constantly changing. This is due to the introduction of new technologies to streamline the flow of drugs across the different entities that take part in the process. There are three main entities when it comes to the overall landscape of the pharmaceutical supply chain industry:

- Drug manufacturers
- Distributors or wholesalers
- Consumers

Drug manufacturers

The pharmaceutical supply chain process starts from the manufacturing of the drug or biologic. They are manufactured by pharmaceutical organizations themselves in their own facilities or are outsourced to manufacturers who manufacture the drugs on their behalf. Drug manufacturers are responsible for ensuring the continuous supply of drugs for their consumers. They track market demand and make adjustments to their manufacturing volumes accordingly. It is also critical they do not over manufacture the drugs as the drugs have a very specific shelf life with a set expiry date. Specialty drug manufacturers, such as the ones that manufacture therapies using biologics in small batches, have very specific requirements around the demand that can be tracked more easily. This is due to the precise nature of the therapy and the circumstances in which it should be used. Drugs that are mass-produced in large batches, however, require more careful data analysis to determine the trends in demand across large groups of consumers. Manufacturers also have a say in the pricing of the prescription drug. The pricing depends on a number of factors, such as the cost of production, the demand for the drug, and the existing competition for the drug. The pricing decides the **wholesale acquisition cost** (**WAC**) of the drug.

Distributors or wholesalers

Drug manufacturers sell their products to consumers such as pharmacies, clinics and healthcare facilities, retailers, and other specialty institutions. In some cases, they may even sell them to health plans and government purchasers. It is difficult for drug manufacturers to track the supply chain demand across these large groups of consumers. As a result, they rely on **wholesalers (distributors)**. Wholesalers act as a bridge between drug manufacturers and consumers. The manufacturers are responsible for distributing the drug to the wholesalers from the manufacturing facilities where the drugs are produced. This is far easier to accomplish as the number of wholesalers that manufacturers have to deal with is far fewer than a large number of direct consumers. The wholesaler acquires the drugs in bulk from manufacturers, manages the inventory of the drugs, and distributes the drugs to the consumers. In return, they receive a fee for their services and are also entitled to a discount due to bulk purchasing.

Consumers

Once the distributors have the drug, they send it to thousands of pharmacies, clinics, or facilities that can make it available to patients who need it. Pharmacies account for more than 70% of consumers in the prescription drug market. Pharmacies purchase the drugs from wholesalers at a pre-negotiated price under a contractual agreement. This price may be different from the WAC and depends on multiple factors, such as the size of the pharmacy chain, the proportion of the cost of the drug that the payers are willing to cover, and the out-of-pocket expenses for the patients. A key player in the middle of all these negotiations is a **pharmacy benefit manager** (PBM). PBMs help reduce the cost of prescription drugs by negotiating better rebates from manufacturers and pharmacies. They encourage the use of low-cost generics versus expensive brand medications and manage the overall prescription drug benefits.

It does seem like a straightforward operation, but the pharmaceutical supply chain industry has been facing multiple challenges lately. Let us summarize some of those in the next section.

Summarizing key challenges faced by the pharmaceutical supply chain industry

With the introduction of new drugs and therapies, unique ways of selling them are becoming more predominant. Specialized drugs and biologics require complex storage infrastructure and facilities to maintain conditions in which the drugs need to be stored. Moreover, there is increased scrutiny of increasing prescription drug prices and the role the pharmaceutical supply chain plays in the price of drugs. Drug manufacturers are required to maintain higher levels of transparency in the entire process to ensure regulators have visibility into details of the manufacturing process. All these factors are responsible for creating more complexities in the pharmaceutical supply chain life cycle. Moreover, manufacturers usually cannot pre-plan the manufacturing of new drugs until they are certain of their success in the clinical trial process. They have to wait for the drug to be cleared for production before they can start setting up the manufacturing facilities. This can cause supply shortages and results in bottlenecks that could cause disruptions in the manufacturing process. Lastly, it is a very competitive landscape, and drug manufacturers need to come up with new manufacturing and distribution strategies all the time to keep up with the competition and maintain profitability. All this introduces a need for regular modernization of supply chain processes.

At the end of the day, a thriving and robust pharmaceutical supply chain is good for us. It helps get much-needed medications to patients in a timely and cost-effective manner. Let us now look into the role that sales representatives play in this overall landscape.

Introducing pharmaceutical sales

The pharmaceutical sales industry is a highly competitive, fast-paced environment that involves selling medical products to consumers who can distribute them among patients. For prescription drugs, it is the physician who is the target consumer for the drug as physicians prescribe drugs to patients. In a situation where there are multiple brands selling drugs for treating the same conditions, physicians can provide an edge to one brand versus the other. To do this, however, is not easy. Drugs that treat common clinical conditions often consist of the same formula, and there are usually generic substitutes available at a far lesser cost. To help, pharmaceutical industries and distributors hire **pharmaceutical sales representatives (pharma sales reps)** who are tasked with increasing sales for a particular brand of drug.

A pharma sales rep is one of the most sought-after jobs in the country and it pays a high salary and commissions. However, it is also very demanding. Unlike other sales reps, pharma sales reps do not sell directly to patients. Instead, they utilize a variety of ways to influence providers to prescribe more of their drugs. There is no magic bullet or manual that helps them succeed at this. In fact, the process may vary based on who the provider is. They may visit a provider multiple times to educate them about the product they are trying to sell. In essence, they need to be knowledgeable about the product and must be able to answer questions that providers might have. They also need to be aware of the competitive landscape for the product, and they must know why their product is better than the competitors'.

The job of a pharma sales rep goes beyond their knowledge and domain experience. It has a lot to do with the rep's networking and relationship-building skills. They need to maintain a long-term relationship with the providers they are targeting and visit them repeatedly to update them about their products. A lot of these visits may not result in any substantial outcome for the pharma company. Instead, the purpose of the visit is to keep the providers up to date about the company's future plans and roadmap, discuss benefits and case studies of the product, address any questions about the products as they come up from patients who are receiving them, and also gather competitive intelligence.

Technologies play a huge role in giving sales reps an edge over their competitors. The sales data can uncover a variety of metrics that help reps improve their performance. It helps them target the right providers and analyze market sentiment to better inform them on the field. Let us now look at how ML can help improve the pharmaceutical supply chain and sales.

Applying ML to the pharmaceutical supply chain and sales

The evolution of the global pharmaceutical industry has introduced new challenges and complexities in the supply chain process. It needs to cater to new distribution mechanisms for therapeutics, adhere to new regulations around supply chain transparency, cater to storage requirements of specialized biologics, and also be flexible enough to handle changes. Moreover, the supply chain should be resilient enough to avoid disruptions. These disruptions can be caused by a variety of reasons such as natural disasters, equipment failures, shortage of raw materials, or even cyber-attacks on IT infrastructure.

ML can help reduce supply chain disruptions and optimize them to reduce waste and increase efficiency. Let us now look at some of the common ways in which we can apply ML to the pharmaceutical supply chain and sales workflows.

Targeting providers and patients

As described earlier, pharmaceutical sales involves visiting providers multiple times to educate them about the benefits of the product. These visits are time-consuming, and going about them randomly can waste a lot of time. Hence, finding the right target audience for the pharmaceutical product is essential.

One of the common ways that ML can help is by identifying probable providers who are more likely to prescribe the product. This can be done using a **recommendation engine**. A recommendation engine is an ML model that recommends the most important or relevant items to a consumer. Recommendation engines have a wide array of applications. For example, they can be used to suggest books to buyers based on their past reading habits, and they can suggest movies to users based on their past likes or dislikes of certain genres or categories. In the case of pharmaceutical sales, they can look at multiple features about the provider—such as their specialty, their past case details, and their patient population demographics—to determine how likely they are to prescribe a certain product. Similarly, recommendation engines can also be applied to patient populations to find cohorts of patients who can have the most positive outcomes from the use of the medical product. In fact, patients can also use them to get recommendations on providers, based on their clinical conditions. To make it easier for customers to create accurate recommendation engines, **Amazon Web Services** (AWS) provides Amazon Personalize, a fully managed service to create personalized recommendations for a variety of use cases. To learn more about Amazon Personalize, visit the following link: `https://aws.amazon.com/personalize/`.

Forecasting drug demand and sales outlook

Demand forecasting is an integral part of the pharmaceutical supply chain industry. It helps route drugs appropriately based on their projected demand, allows distributors to keep adequate stock, helps replenish raw materials based on expected spikes in demand, and also helps determine projected profits and revenue for a product. Forecasts are generated using a temporal attribute that stores the time feature. It defines the granularity of the forecast, such as yearly, monthly, and weekly. To generate a forecast, you need a dataset that has that temporal feature and an associated value you want to predict.

For example, to forecast the monthly demand for a drug, you need data from past months that have the monthly sales quantities for the drug. It may require that you derive this dataset from raw transactional data in a sales database if it's not readily available. In many cases, the forecast can be improved if we add more associated features to the dataset. These associated features help provide more context to the forecasting engine by adding features to the training data. Another way forecasts can become more complex is by adding an associated time series feature. This time series is not the attribute we are predicting but provides additional information about the forecast training dataset.

Forecasting algorithms utilize a range of simple statistical methods to complex **deep learning (DL) neural networks (NNs)** that understand the context between the target time series, additional features, and associated time series. The more complex the forecasting problem is, the more complex the feature engineering and modeling steps. To make it easy, AWS provides Amazon Forecast, a fully managed service to generate accurate forecasts from a variety of time series datasets. We will introduce Amazon Forecast in the next section and use it to build a pharmaceutical sales forecasting model.

Implementing predictive maintenance

Drug manufacturing equipment needs to be serviced and cared for regularly to avoid failures. A shutdown of the equipment can cause major disruptions in the drug supply chain. Moreover, since a lot of the supply chain steps are sequential, the impact of failures in the manufacturing facility can extend far beyond the facility itself and impact downstream drug distributions. A shutdown of a facility that manufactures drugs can cause major losses for the pharmaceutical company. Manufacturers can use ML models to carry out predictive maintenance of manufacturing equipment and prevent total shutdowns.

Sensors mounted on the equipment collect data at regular intervals. These datasets are then aggregated to look at trends around equipment performance. Models trained to predict the need for maintenance can then sound alarms and notifications in case maintenance is needed. If the equipment cannot be connected to a network for some reason, these models can be deployed directly on the equipment to generate real-time inference on equipment data. ML models can also help reduce manual errors in manufacturing processing by automating repeatable tasks. For example, computer vision algorithms can identify manufacturing defects using images from the manufacturing facility. The use of advanced analytics and ML in manufacturing facilities helps modernize manufacturing practices and is the basis of **smart factories** with an interconnected network of machines.

Understanding market sentiment and completive analysis

Market sentiment, taken from reviews about a drug, is an essential factor in predicting how successful it will be in the market. A bad review on a credible forum can cause a negative impact on sales of the product. Moreover, if the review contains information about serious side effects due to the drug, it may cause further scrutiny for the pharmaceutical company. While you cannot prevent bad reviews, it is still important to track them to understand the market sentiment around the drug. For example, if there are multiple reviews about a side effect from the drug that was not recorded in the trials, the scientists can proactively evaluate the drug for side effects and issue a warning if necessary. If, however, the side effect is a false alarm, you can use marketing channels and educational material to clear the air. This creates a proactive way of engaging your customer base rather than being reactive, whereby you wait for bad reviews to negatively impact the sales of the drug and then take corrective action. Another important aspect of market analysis is to understand the competitive landscape for the drug. For example, reviews that compare one drug to another might be from competitors, and the details in the comparison could be important to differentiate one product from another.

Natural language processing (**NLP**) algorithms can detect sentiments (positive, negative, neutral) from free text notes and associate them with the drug. They can also extract important information from reviews using **named entity recognition** (**NER**) and classify the reviews based on multiple categories. These techniques help reduce human dependency by reading these reviews and determining the ones that are of interest.

These are just some of the common applications of ML in the pharmaceutical supply chain. Over the years, as the supply chain industry is transformed and ML technology becomes more accessible, the applications of this technology in the pharmaceutical industry are only expected to grow. Now that we have a good understanding of the pharmaceutical supply chain and sales industry, let us dive into Amazon Forecast and use it to build a forecasting model for pharmaceutical sales.

Introducing Amazon Forecast

Amazon Forecast is a managed AI service from AWS that allows you to create accurate time series forecasting models. The service allows you to import time series datasets, preprocess and clean up the datasets for training, train a forecasting model, and deploy it to generate forecasts on future time intervals. You can then export these forecasts to visualize the results on reporting platforms such as **Amazon QuickSight**. You can also visualize the results on the AWS console or use the API to get real-time results.

Amazon Forecast provides multiple options when it comes to the choice of algorithms to use while training a model. They range from statistical methods to DL algorithms, and you can choose them based on factors such as the volume of the data, the number of missing values in the data, and also the complexity of the dataset. Let us look in more detail at the algorithms that Amazon Forecast provides.

Amazon Forecast algorithms

The models created by Amazon Forecast can use one of the following algorithms for training:

> **Note**
> You also have the choice of letting Forecast choose the optimal algorithm for you by choosing the **AutoML** option. This option lets Forecast choose the optimal algorithm to train the model.

- **Autoregressive Integrated Moving Average (ARIMA)**: ARIMA is a statistical algorithm commonly used for time series forecasting. It learns by the extrapolation of seasonal oscillations in time series data and filtering out any noise that does not correlate to the seasonal patterns. It uses these patterns to predict future values in the time series. To learn more about ARIMA, look at the following link: `https://en.wikipedia.org/wiki/Autoregressive_integrated_moving_average`.

- **Exponential Smoothing (ETS)**: ETS is another common statistical method used for time series forecasting. It computes a weighted average across all data points in the input time series. These weights decrease over time and are based on a smoothing parameter that remains constant. To learn more about ETS, look at the following link: `https://en.wikipedia.org/wiki/Exponential_smoothing`.

- **Non-Parametric Time Series (NPTS)**: The NTPS algorithm is based on probabilistic methods and is especially useful for sparse time series with 0 values that occur in bursts. It uses past observations to predict future value distributions of a given time series. The time series in the dataset can be of different lengths as NTPS computes predictions for each time series individually. The NTPS algorithm has the following variants:

  - **NPTS**: This variant assumes that recent values in the time series have more weightage than values in the distant past. As a result, it associates weights based on how far they are from the current time. Predictions are generated using all observations in the time series.

  - **Seasonal NPTS**: This variant looks at the value from past seasons instead of looking at all observations. For example, for a daily forecast that predicts the value on day d, the algorithm learns by looking at the observations on day d in the past weeks. The weights become lesser as the observations move further into the past.

  - **Climatological forecaster**: This variant of NPTS samples all observations from the past with uniform probability.

  - **Seasonal climatological forecaster**: This variant of NPTS samples seasonal observations from the past with uniform probability.

- **Prophet**: Prophet is a **Bayesian** structural time series algorithm that is especially useful on datasets with extended time periods of detailed observations. It uses an additive regression model with a piecewise or logistic growth curve trend and is able to handle missing values or large outliners in the training dataset. For more details on Prophet, check out the following link: `https://research.facebook.com/blog/2017/2/prophet-forecasting-at-scale/`.

- **Convolutional Neural Network – Quantile Regression (CNN-QR)**: CNN-QR is a proprietary model that can be used to predict one-dimensional time series using causal **convolutional NNs** (**CNNs**). It trains one global model from a large collection of time series and uses a quantile decoder to generate predictions. Because of its NN-based architecture, CNN-QR can generate forecasts with large and complex datasets. It is also useful for cold-start scenarios where there is little or no existing historical data. This algorithm is also computationally more expensive compared to other algorithms that are based on statistical methods.

- **DeepAR+**: DeepAR+ is a **supervised learning** (SL) algorithm based on **recurrent NN (RNN)**-based architecture. The algorithm trains a single global model based on hundreds of time series in your dataset. DeepAR+ can accept associated feature time series in addition to the target time series. Moreover, each target time series can have a number of categorical features. These categorical features allow the algorithm to learn the typical behavior of the target within a group. The algorithm also generates computed features based on the granularity of the target time series. For example, for an hourly forecast, the algorithm generates hour-of-day, day-of-week, day-of-month, and day-of-year features. Because of its NN-based architecture, the algorithm is suitable for large complex datasets with many associated features and time series. It is also computationally more expensive than other algorithms based on statistical methods.

As you can see, Amazon Forecast provides a variety of algorithms for different forecasting problems. To know more about these algorithms, you can visit the following link: `https://docs.aws.amazon.com/forecast/latest/dg/aws-forecast-choosing-recipes.html`.

Let us now take a look at the various steps you need to perform to generate a forecast using the Amazon Forecast service.

Importing a dataset

The first step in using Amazon Forecast is to create and import a dataset. The dataset is defined by its **schema**. During the import job, the forecast service assumes the exact same structure of the dataset as defined in the schema. The dataset is organized as a group within Forecast known as a **dataset group**, which consists of associated information in the form of complementary dataset files. Each dataset group consists of a **target time series**, **related time series**, and **item metadata**. The target time series file is the minimum required file for the forecast dataset and consists of the field that we want to predict in our forecast. The related time series dataset is a time series that is associated with the target time series, but it doesn't have the target field in it. Lastly, the item metadata doesn't consist of any time series but has fields related to the target. Note that both the related time series and item metadata are optional within a dataset group. Moreover, only DeepAR+ and CNN-QR algorithms can make use of the item metadata dataset.

Amazon Forecast also provides pre-defined dataset domains for forecasting problems in specific domains. These include domains such as retail, inventory planning, and workforce. It also provides a custom domain for any type of time series that does not fall into the pre-defined domain. Once a dataset and dataset group are created, you can train a predictor using that dataset.

Training forecasting models

The next step is to train a prediction model using our dataset. A key part of this step is the choice of the forecasting algorithm. As described earlier, Amazon Forecast provides a range of statistical and NN-based models for generating predictions. It is important to consider the choice of algorithm carefully and make sure you apply the right algorithm to the dataset and problem at hand. You may also choose the **AutoML** option to allow Forecast to choose the right algorithm for you based on the dataset, as mentioned previously. One great way to compare the accuracies of your own chosen algorithm and the auto predictor is to upgrade your own predictor to an auto predictor. This creates a new version of the trained model so that you can run comparisons against both versions of the model. To train a predictor, Amazon Forecast needs the dataset group, the forecast frequency (hourly, daily, weekly, and so on), and the forecast horizon, which is the number of time steps to generate in the forecast.

Amazon Forecast generates prediction metrics such as **Root Mean Square Error** (**RMSE**) and **Weighted Quantile Loss** (**wQL**) to evaluate the model performance. You may also choose to retrain your models when new data is available and monitor the performance of your model using the metrics generated by Forecast or using the built-in predictor monitoring feature. When you do this, the original predictor remains intact and a new predictor with a new **Amazon Resource Name** (**ARN**) is created. Once you have trained your prediction model, it is time to generate forecasts from it.

Generating forecasts

Forecast generation corresponds to the inference step in a typical ML pipeline. Amazon Forecast deploys the trained model as a predictor that is ready to be queried to generate forecasts. A forecast is generated for an item in the dataset. Once generated, you can query the forecast using the API provided by the service; while querying, you can define filter criteria as key-value pairs in the query corresponding to feature-value pairs in the dataset. You can also export the forecast results as a CSV file for integrating it with other services from AWS or for integration with other downstream systems. Lastly, you can explain the model predictions using the expandability features of Amazon Forecast. This allows you to generate impact scores for features that contribute to the model results.

There are other features in Amazon Forecast such as what-if analysis, which lets you mock up different scenarios about how a forecast may be impacted if you change the values of certain input features in the training data. It is done using a related time series dataset and models how changes to the baseline-related time series dataset will impact the forecast. To know more about all features and capabilities of Amazon Forecast, you can refer to the developer guide here: `https://docs.aws.amazon.com/forecast/latest/dg/what-is-forecast.html`.

Now that we understand the functionalities of Amazon Forecast, we will use this service to generate a time series forecast for predicting pharmaceutical sales in the next section.

Building a pharmaceutical sales forecasting model using Amazon Forecast

In this section, we will use Amazon Forecast to build a forecasting model to predict sales quantities for a drug. We will generate two different forecasts. The first one will be at the hourly granularity, and the second one will be at the daily granularity. We will then visualize the forecasts using a chart. Let's begin by downloading the dataset.

Acquiring the dataset

The dataset for this exercise is available at the following link: `https://www.kaggle.com/code/milanzdravkovic/pharma-sales-data-analysis-and-forecasting/data`.

You can see the contents of the `saleshourly.csv` file here:

Download the files saleshourly.csv and salesdaily.csv from the above portal. Let us look at the contents of the two files in a bit more detail.datum	M01AB	M01AE	N02BA	N02BE	N05B	N05C	R03	R06	Year	Month	Hour	Weekday Name
1/2/14 8:00	0	0.67	0.4	2	0	0	0	1	2014	1	8	Thursday
1/2/14 9:00	0	0	1	0	2	0	0	0	2014	1	9	Thursday
1/2/14 10:00	0	0	0	3	2	0	0	0	2014	1	10	Thursday
1/2/14 11:00	0	0	0	2	1	0	0	0	2014	1	11	Thursday
1/2/14 12:00	0	2	0	5	2	0	0	0	2014	1	12	Thursday

Figure 10.1 – Contents of the saleshourly.csv file

The `saleshourly.csv` file consists of quantities of drugs sold at every hour. The `datum` column consists of timestamps corresponding to each hour. The `M01AB` to `R06` column headings denote the names of the drugs. It also consists of `Year`, `Month`, `Hour`, and `Weekday Name` values for each hour the drug was sold.

Here, you can see the contents of the `salesdaily.csv` file:

datum	M01AB	M01AE	N02BA	N02BE	N05B	N05C	R03	R06	Year	Month	Hour	Weekday Name
1/2/14	0	3.67	3.4	32.4	7	0	0	2	2014	1	248	Thursday
1/3/14	8	4	4.4	50.6	16	0	20	4	2014	1	276	Friday
1/4/14	2	1	6.5	61.85	10	0	9	1	2014	1	276	Saturday
1/5/14	4	3	7	41.1	8	0	3	0	2014	1	276	Sunday
1/6/14	5	1	4.5	21.7	16	2	6	2	2014	1	276	Monday

Figure 10.2 – Contents of the salesdaily.csv file

The `salesdaily.csv` file consists of quantities of drugs sold every day. The `datum` column consists of the date when the drug was sold. The `M01AB` to `R06` column headings denote the names of the drugs. It also consists of `Year`, `Month`, `Hour`, and `Weekday Name` values for each day the drug was sold.

We will use these two files to create two forecasting models to forecast the sale of a drug. The first one will be an hourly forecast, and the second one will be a daily forecast. To complete the exercise, proceed to execute the Jupyter notebook by following the steps in the next section.

Running the Jupyter notebook

The notebook for this exercise is saved on GitHub here: `https://github.com/PacktPublishing/Applied-Machine-Learning-for-Healthcare-and-Life-Sciences-using-AWS/blob/main/chapter-10/pharma_sales_forecasting.ipynb`.

The repository was cloned as part of the steps in the *Technical requirements* section. You can access the notebook from GitHub by following these steps:

1. Open the SageMaker Studio interface.

2. Navigate to the `Applied-Machine-Learning-for-Healthcare-and-Life-Sciences-using-AWS/chapter-10/` path. You should see a file named `pharma_sales_forecasting.ipynb`.

3. Select **New Folder** on the top of the folder navigation pane in SageMaker Studio, as shown in the following screenshot:

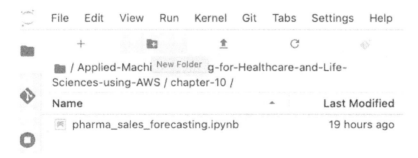

Figure 10.3 – SageMaker Studio UI showing the New Folder button

4. Name the folder `data`. Next, click the **Upload Files** icon on the top of the navigation pane in SageMaker Studio, as shown in the following screenshot:

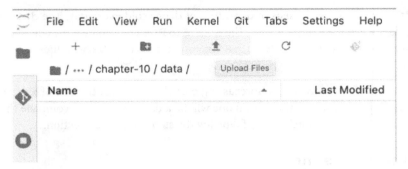

Figure 10.4 – SageMaker Studio UI showing the Upload Files button

5. Upload the `saleshourly.csv` and `salesdaily.csv` files that you downloaded previously to the `data` folder.

6. Go back one folder to `Chapter-10` and click on the `pharma_sales_forecasting.ipynb` file. This will open the Jupyter notebook. Follow the instructions in the notebook to complete the exercise.

At the end of the exercise, we will have generated forecasts from two models for hourly and daily intervals.

This concludes our exercise. Please make sure you stop or delete your Amazon Forecast resources to avoid incurring charges, as described in the following link: `https://docs.aws.amazon.com/forecast/latest/dg/manage-resources.html`.

Summary

In this chapter, we learned about the pharmaceutical supply chain industry, the different entities involved, their roles, and the challenges that the industry faces today. We also learned about pharmaceutical sales and how pharma sales reps work to increase awareness and knowledge about their products. We looked at how ML is influencing the pharmaceutical supply chain industry by looking into common applications of ML in pharmaceutical supply chain and sales. We then learned about Amazon Forecast and its various features to create simple and complex forecasting models. Lastly, we used Amazon Forecast to build a pharmaceutical sales forecasting model.

In *Chapter 11, Understanding Common Industry Challenges and Solutions*, we will get an understanding of the challenges that the healthcare and life sciences industry faces when utilizing ML. We will also look at how to solve some of those challenges and adhere to regulatory requirements in the healthcare and life sciences industry.

Part 4:
Challenges and the Future of AI in Healthcare and Life Sciences

In the final part of the book, we will examine some commonplace challenges in the industry and look ahead to what the future may hold.

- *Chapter 11, Understanding Common Industry Challenges and Solutions*
- *Chapter 12, Understanding Current Industry Trends and Future Applications*

Understanding Common Industry Challenges and Solutions

The use of ML in healthcare and life sciences can have a transformational impact on the industry. It can help find a cure for incurable diseases, prevent the next pandemic, and help healthcare professionals become more efficient at providing care. All of this helps uplift an entire population with a better quality of life and timely care. However, there are some challenges in applying ML to the healthcare and life sciences industry. Healthcare data is extremely private, requiring the highest levels of security. This prevents open sharing and collaboration among researchers to advance this field, getting in the way of innovation. There are also a variety of regulations in place that make it difficult for practical ML applications to be readily used in mission-critical workloads in the industry. Moreover, as ML models become better, they also become more complex. **Neural networks** (**NNs**) that have known to be **state-of-the-art** (**SOTA**) are usually made of extremely complicated architectures and may contain billions of parameters. This makes it extremely difficult to understand why the model is predicting the way it is. Without a good understanding of this, we are looking at a "**black box**" that predicts an output without ensuring whether it's fair or transparent. This is a deterrent for adoption in the healthcare and life sciences industry, which promotes transparency and reporting to regulatory bodies.

However, not all is lost. While there are still many challenges, there are a good number of success stories around the application of ML to healthcare and life sciences workloads. For example, we have seen both Pfizer-BioNTech and Moderna develop an entirely new vaccine for the novel coronavirus in record time by applying ML to different stages of the drug discovery and clinical trial pipeline. We have seen companies such as Philips and GE Healthcare that have revolutionized medical imaging by using **computer vision** (**CV**) models on their equipment. We also see numerous examples of start-ups being formed with the goal of curing cancer by utilizing ML models that can find a variety of characteristics from genomic and proteomic data. All these examples prove that there are ways in which you can apply ML technology successfully in this industry. This is happening by utilizing workarounds and solutions that enable you to address and meet these challenges head-on. It is a matter of knowing how to address them and finding the best alternative by balancing complexity, transparency, and usefulness.

In this chapter, we will look at some of the common challenges that come in the way of applying ML to workloads in the healthcare and life sciences industry. We will then look at some ways to address those challenges by understanding workarounds and solutions. We will get introduced to new features of SageMaker, **SageMaker Clarify** and **SageMaker Model Monitor**, that help you create fair and transparent AI models for healthcare and life sciences workloads. Lastly, we will use this feature to get a better understanding of an ML model that predicts healthcare coverage amounts for patients across different demographics.

Here are the main topics we will cover in this chapter:

- Understanding challenges with implementing ML in healthcare and life sciences
- Understanding solution options for solving these challenges
- Introducing SageMaker Clarify and Model Monitor
- Detecting bias and explaining model predictions for healthcare coverage amounts
- Viewing bias and explainability reports in SageMaker Studio

Technical requirements

You need to complete the following technical requirements before building the example implementation at the end of this chapter:

1. Complete the steps to set up the prerequisites for Amazon SageMaker, as described here: `https://docs.aws.amazon.com/sagemaker/latest/dg/gs-set-up.html`.

2. Onboard to SageMaker Domain using **Quick setup**, as described here: `https://docs.aws.amazon.com/sagemaker/latest/dg/onboard-quick-start.html`.

> **Note**
>
> If you have already onboarded a SageMaker Domain from a previous exercise, you do not need to perform *step 2* again.

3. Once you are in the SageMaker Studio interface, click on **File | New | Terminal**.

4. Once in the terminal, type the following commands:

```
git clone https://github.com/PacktPublishing/Applied-
Machine-Learning-for-Healthcare-and-Life-Sciences-using-
AWS.git
```

You should now see a folder named `Applied-Machine-Learning-for-Healthcare-and-Life-Sciences-using-AWS`.

> **Note**
>
> If you have already cloned the repository in a previous exercise, you should already have this folder. You do not need to do *step 4* again.

5. Familiarize yourself with the SageMaker Studio UI components, as described here: `https://docs.aws.amazon.com/sagemaker/latest/dg/studio-ui.html`.

Once you have completed these steps, you should be all set to execute the steps in the example implementation in the last section of this chapter.

Understanding challenges with implementing ML in healthcare and life sciences

We have seen multiple examples of the use of ML in healthcare and life sciences. These include use cases for providers, payers, genomics, drug discovery, and many more. While we have shown how ML can solve some of the biggest challenges that the healthcare industry is facing, implementing it at scale for healthcare and life sciences workloads has some inherent challenges. Let us now review some of those challenges in more detail.

Healthcare and life sciences regulations

Healthcare and life sciences is a highly regulated industry. There are laws that protect a patient's health information and ensure the security and privacy of healthcare systems. There are some laws that are specific to countries that the patients reside in, and any entity that interacts with data for those patients needs to comply with those rules. Let's look at some examples of such regulations:

*   **Health Insurance Portability and Accountability Act (HIPAA)**: In the US, the **Department of Health and Human Services (HHS)** created a set of privacy rules under HIPAA that standardized the protection of personally identifiable health information also known as **protected health information (PHI)**. As part of these standards, it included provisions to allow individuals to understand and control how their health information is used. This applies to all covered entities as defined in HIPAA. They include healthcare providers, health plans, healthcare clearinghouses, and business associates.

*   **Health Information Technology for Economic and Clinical Health (HITECH)**: HITECH is another regulation specific to US healthcare. HITECH was created to increase the adoption of the **electronic health record (EHR)** system. It includes provisions for grants and training for employees who are required to support health IT infrastructure. It also includes requirements for data breach notifications and increased penalties for HIPAA violations. As a matter of fact,

HIPAA and HITECH are closely related in the way they reinforce each other. For example, under HITECH, the system requirements that were put in place ensure that they do not violate any HIPAA laws.

- **General Data Protection Regulation (GDPR)**: GDPR is a European law that was implemented in 2018. It intends to harmonize data protection and privacy laws across the **European Union (EU)**. It applies to any organization (operating within or outside of the EU) that processes and accesses the personal data (**personally identifiable information**, or **PII**) of anyone in the EU and aims to create consistency across EU member states on how personal data can be processed, used, and exchanged securely.

- **GxP**: GxP compliance broadly refers to **good practices** for life sciences organizations that manufacture medical products such as drugs, medical devices, and medical software applications. It can have variations depending on the area of life sciences that it's applied in. For example, you can have **Good Manufacturing Practices (GMPs)**, **Good Laboratory Practices (GLPs)**, and **Good Clinical Practices (GCPs)**. The goal of all GxP compliance variations is to ensure safety and quality through the use of validated systems that operate per their intended use.

Let us now look at some security- and privacy-related requirements when it comes to healthcare and life sciences workloads.

Security and privacy

Healthcare data needs to be handled with care to ensure there are no violations of any of the regulatory guidelines such as **HIPAA** or **GDPR**. There are huge penalties for organizations that do not comply. For example, the total fines for healthcare organizations that reported HIPAA violations between 2017 and 2020 were approximately USD $80 million. Moreover, the laws require you to report violations of data breaches in a timely manner to the public. This creates further embarrassment for the organizations concerned and leads to a lack of trust and negative consumer sentiment. There is no doubt that it is absolutely essential for healthcare organizations to adhere to these regulations by keeping security and privacy as an integral part of their architecture. The rules are carefully thought out and created to protect patients and keep them and their data safe.

ML is no exception when it comes to implementing security and privacy. ML data needs to be properly secured and anonymized. It needs to be access-controlled so that only authorized people or systems have access to it. The environment in which the model is trained and hosted needs to have appropriate security controls in place so that it cannot be breached. All of these are essential steps that may sometimes slow down the pace of the adoption of ML in healthcare. Nonetheless, the risk of not having these measures in place outweighs the reward that any innovation could bring.

Bias and transparency in ML

Another challenge for implementing ML technology in healthcare and life sciences workflows is the emphasis on fairness. A model that predicts a positive outcome for one demographic more often than another is said to be **biased** toward that demographic and cannot be considered fair. In healthcare, a biased model can be detrimental in a variety of scenarios. For example, a claims adjudication model can be biased toward approving claims for a certain gender or ethnicity more often than others. This provides better healthcare coverage for a certain group of people compared to others. We are all working toward creating a world where health equity is prioritized. This means healthcare should be available to all and should provide equal benefits for everyone. An unfair model goes against this principle. As a result, when building models for healthcare use cases, data scientists should take into account the necessary steps that ensure fairness in model predictions.

Reproducibility and generalizability of ML models

Another challenge for the adoption of ML in healthcare is the concept of **generalizability**. This essentially underscores the importance of the model results being **reproducible** across different samples of real-world data. It happens very often that we see a model that performs really well in research environments on specific test datasets but does not retain the same performance in real-world scenarios. This may be for a variety of reasons. One common reason is that the data on which the model is trained and evaluated does not mimic real-world behavior. It may also be biased toward one group or class.

Now that we have an understanding of some of the common challenges that organizations face in the adoption of ML in healthcare and life sciences workflows, let us dive into some solutions that may help address these challenges.

Understanding options for solving these challenges

While there are known challenges, there are ways organizations can successfully adopt ML in healthcare and life sciences workloads. One important factor to consider is at what stage in the workflow you apply ML. While there have been improvements to the technology, it is not a replacement for a trained medical professional. It cannot replace the practical experience and knowledge gathered through years of practice and training. However, technology can help make medical professionals more efficient at what they do. For example, instead of relying on an ML model to make diagnostic decisions for a patient, it can be used to recommend a diagnosis to a healthcare provider and let them make a final decision. It can act as a tool for them to search through a variety of case reports and medical history information that will help them in their decision-making process. The medical professional is still in control while getting a superpower that has the ability to automate mundane and repeatable tasks and also augment the practical experience of a provider instead of replacing it. Let us now look at some common technical methods to consider in your ML pipeline to make it suitable for healthcare and life sciences workloads.

Encrypting and anonymizing your data

Data encryption is a common way to protect sensitive information. It involves transforming your original data into another form or code that is impossible to understand or decipher. The only way to retrieve the original plain text information is via **decryption**, which reverses the operation. To achieve this, you must have access to the **encryption/decryption key**, which is a unique **secret key**. No one without access to this secret key can access the underlying data, hence making it useless.

Utilizing encryption algorithms to transform your sensitive healthcare information helps keep the data safe. The customers need to just keep the secret key secure instead of a whole database, which makes the process much more manageable and reduces the risk involved. There are some things to keep in mind, though. Firstly, using these algorithms involves extra processing, both while encrypting and decrypting. This introduces more compute overhead on your systems. Applications that rely on reading and displaying encrypted data may perform sluggishly and impact user experience. Moreover, there is a risk of hackers breaking encryption keys and getting access to sensitive information if they are too simple. Hackers rely on trial and error and run software that can generate multiple permutations and combinations until they find the right one. This pushes organizations to create more complicated and lengthy keys, which in turn adds to the processing overhead. Therefore, it is important to come up with the correct encryption setup for your data, both in transit and at rest. Understand the compute overhead, plan for it, and ensure you have enough capacity for the types of queries you run on your encrypted data.

Another way to de-sensitize your data is by using **anonymization**. Though encryption may be considered a type of anonymization technique, data anonymization involves methods to erase sensitive fields containing personally identifiable data from your datasets. One common way of data anonymization is **data masking**, which involves removing or masking sensitive fields from your dataset. You can also generate synthetic versions of the original data by utilizing statistical models that can generate a representative set similar to the original dataset. The obvious consideration here is the method of anonymization. While methods such as data masking and erasing are easy to implement, they may remove important information from the dataset that can be critical to the performance of the downstream analytics or ML models. Moreover, removing all sensitive information from the data affects its usability for **personalization** or **recommendations**. This can affect end-user experience and engagement. Security should always be the top priority for any healthcare organization, and striking a balance between technical complexity and user experience and convenience is a key aspect when deciding between data anonymization and encryption.

Explainability and bias detection

The next consideration for implementing ML models in healthcare workflows is how open and transparent the underlying algorithm is. Are we able to understand why a model is predicting a particular output? Do we have control over the parameters to influence its prediction when we choose to do so? These are the questions that model **explainability** attempts to answer. The key to explaining model behavior is to **interpret** it. Model interpretation helps increase trust in the model and is especially critical in healthcare and life sciences. It also helps us understand whether the model is biased toward a particular class or demographic, enabling us to take corrective action when needed. This can be due to a bias in the training dataset or in the trained model. The more complicated a model is, the more difficult it is to interpret and explain its output. For example, in NNs that consist of multiple interconnected layers that transform the input data, it is difficult to understand how the input data impacts the final prediction. However, a single-layer regression model is far easier to understand and explain. Hence, it is important to balance SOTA with simplicity. It is not always necessary to use a complicated NN if you can achieve acceptable levels of performance from a simple model. Similarly, it may not always be necessary to explain model behavior if the use case it's being implemented for is not mission-critical.

Model explanations can be generated at a local or a global level. **Local explanations** attempt to explain the prediction from a model for a single instance in the data. A way to generate a local explanation for your model is to utilize **Local Interpretable Model-Agnostic Explanations** (**LIME**). LIME provides local explanations for your model by building surrogate models using permutations of feature values of a single dataset and observing its overall impact on the model. To know more about LIME, visit the following link: `https://homes.cs.washington.edu/~marcotcr/blog/lime/`. Another common technique for generating local explanations is by using **SHapley Additive exPlanations** (**SHAP**). The method relies on the computation of Shapley values that represents the feature contribution for each feature toward a given prediction. To learn more about SHAP, visit the following link: `https://shap.readthedocs.io/en/latest/index.html`. **Global explanations** are useful to understand the model behavior on the entire dataset to get an overall view of the factors that influence model prediction. SHAP can also be used to explain model predictions at a global level by computing feature importance values over the entire dataset and taking an average for it. Or, you can use a technique called **permutation importance**. This method selectively replaces features with noise and computes the model performance to adjudicate how much impact the particular feature has in overall predictions.

Building reproducible ML pipelines

As ML gets more integrated with regular production workflows, it is important to get consistent results from the model under similar input conditions. However, model outputs may change over a period of time as the model and data evolve. This could be for a variety of reasons, such as a change in business rules that influence real-world data. This is commonly referred to as **model drift** in ML. As you may recall, ML is an iterative process, and the model results are a probabilistic determination of the most likely outcome. This introduces a new problem of reproducibility in ML. A model or a pipeline is said to be reproducible if its results can be recreated with the given assumptions and features. It is an important factor for **GxP compliance**, which requires validating a system for its intended use. There are various factors that may influence inconsistent model outputs; it mostly has to do with how the model is trained. Retraining a model with a different dataset or a different set of features will most likely influence the model output. It can also be influenced by your choice of training algorithm and the associated hyperparameters.

Versioning is a key factor in model reproducibility. Versioning each step of the ML pipeline will help ensure there are no variations between them. For example, the training data, feature engineering script, training script, model, and deployed endpoint should all be versioned for each variation. You should be able to associate a version of the training dataset with a version of the feature engineering script, a version of a training script, and so on. You should then version this entire setup as a pipeline. The pipeline allows you to create traceability between each step in the pipeline and maintain it as a separate artifact. Another aspect that could influence model output variation is the compute environment on which the pipeline is run. Depending on the level of control you need on this aspect, you may choose to create a **virtual environment** or a **container** for running your ML model pipeline. By utilizing these methods, you should be able to create a traceable and reproducible ML model pipeline that runs in a controlled environment, reducing the chances of inconsistency across multiple model outputs.

Auditability and review

Auditability is key for a transparent ML model pipeline. An ideal pipeline logs each step that might need to be analyzed. Auditing may be needed to fulfill regulatory requirements, or it may also be needed to debug an issue in the model pipeline. For auditing to occur, you need to ensure that there is centralized logging set up for your ML pipeline. This logging platform is like a central hub where model pipelines can send logs. The logging platform then aggregates these logs and makes them available for end users who have permission to view them. A sophisticated logging platform also supports monitoring of real-time events and generates alerts when a certain condition is fulfilled. Using logging and monitoring together helps prevent or fix issues with your model pipeline. Moreover, the logging platform is able to aggregate logs from multiple jobs and provide log analytics for further insights across your entire ML environment. You can automatically generate audit reports for your versioned pipeline and submit them for regulatory reviews. This saves precious time for both the regulators and your organization.

While logging and monitoring are absolutely essential for establishing trust and transparency in ML pipelines, they may sometimes not be enough by themselves. Sometimes, you do need the help of humans to validate model output and override those predictions if needed. In ML, this is referred to as a **human-in-the-loop** pipeline. It essentially involves sending model predictions to humans with subject-matter expertise to determine whether the model is predicting correctly. If not, they can override the model predictions with the correct output before sending the output downstream. This intervention by humans can also be enabled for selected predictions. Models can generate a confidence score for each prediction that can be used as a threshold for human intervention. Predictions with high confidence scores do not require validation by humans and can be trusted. The ones with confidence scores below a certain threshold can be selectively routed to humans for validation. This ensures only valid cases are being sent for human review instead of increasing human dependency on all predictions. As more and more data is generated with validated output, it can be fed back to the training job to train a new version of the model. Over multiple training cycles, the model can learn the new labels and begin predicting more accurately, requiring even lesser dependency on humans. This type of construct in ML is known as an active learning pipeline.

These are some common solutions available to you when you think about utilizing ML in healthcare and life sciences workflows. While there is no single solution that fits all use cases and requirements, utilizing the options discussed in this section will help ensure that you are following all the best practices in your model pipeline. Let us now learn about two new features of SageMaker, SageMaker Clarify and Model Monitor, and see how they can help you detect bias, explain model results, and monitor models deployed in production.

Introducing SageMaker Clarify and Model Monitor

SageMaker Clarify provides you with greater visibility into your data and the model it is used to train. Before we begin using a dataset for training, it is important to understand whether there is any bias in the dataset. A biased dataset can influence the prediction behavior of the model. For instance, if a model is trained on a dataset that only has records for older individuals, it will be less accurate when applied to predict outcomes for younger individuals. SageMaker Clarify also allows you to explain your model by showing you reports of which attributes are influencing the model's prediction behavior. Once the model is deployed, it needs to be monitored for changes in behavior over time as real-world data changes. This is done using SageMaker Model Monitor, which alerts you if there is a shift in the feature importance in the real-world data.

Let us now look at SageMaker Clarify and Model Monitor in more detail.

Detecting bias using SageMaker Clarify

SageMaker Clarify uses a special processing container to compute bias in your dataset before you train your model with it. These are called **pre-training bias metrics**. It also computes bias metrics in the model predictions after training. These are known as **post-training bias metrics**. Both of these metrics are computed using SageMaker Processing jobs. You have the option to use Spark to execute processing jobs, which is especially useful for large datasets. SageMaker Clarify automatically uses Spark-based distributed computing to execute processing jobs when you provide an instance count greater than 1. The processing job makes use of the SageMaker Clarify container and takes specific configuration inputs to compute these metrics. The reported metrics can then be visualized on the SageMaker Studio interface or downloaded locally for post-processing or integration with third-party applications. Let us now look at some of the common pre-training and post-training bias metrics that SageMaker Clarify computes for you.

Pre-training bias metrics

Pre-training bias metrics attempt to understand the data distributions for all features in the training data and how true they are to the real world. Here are some examples of metrics that help us determine this:

- **Class Imbalance (CI)**: CI occurs when a certain category is over- or under-represented in the training dataset. An imbalanced dataset may result in model predictions being biased toward the over-expressed class. This is not desirable, especially if it doesn't represent real-world data distribution.

- **Difference in Proportions in Labels (DPL)**: This metric tries to determine whether the labels of target classes are properly distributed among particular groups in the training dataset. In the case of binary classification, it determines the proportion of positive labels in one group compared to the proportion of positive labels in another group. It helps us measure whether there is an equal distribution of the positive class within each group in the training dataset.

- **Kullback-Leibler (KL) divergence:** The KL divergence metric measures how different the probability distribution of labels of one class is from another. It's a measure of distance (divergence) between one probability distribution and another.

- **Conditional Demographic Disparity in Labels (CDDL)**: This metric determines the disparity in label distribution within subgroups of training data. It measures whether the positive class and negative class are equally distributed within each subgroup. This helps us determine whether one subgroup (say, people belonging to an age group of 20–30 years) has more positive classes (say, hospital readmissions) than another subgroup (people belonging to a different age group).

These are just four of the most common pre-training metrics that SageMaker Clarify computes for you. For a full list of pre-training metrics, refer to the white paper at the following link: `https://pages.awscloud.com/rs/112-TZM-766/images/Amazon.AI.Fairness.and.Explainability.Whitepaper.pdf`.

Post-training bias metrics

SageMaker Clarify provides 11 post-training bias metrics for models and data that help you determine whether the model predictions are fair and representative of real-world outcomes. Users have a choice to use a subset of these metrics depending on the use case they are trying to implement. Moreover, the definition of fairness for a model also differs from use case to use case, and so does the computation of metrics that determine the concept of fairness for that particular model. In a lot of cases, human intervention and stakeholder involvement are needed to ensure correct metrics are being tracked to measure model fairness and bias. Let us now look at some of the common post-training metrics that SageMaker Clarify can compute for us:

- **Difference in Positive Proportions in Predicted Labels (DPPL)**: This metric measures the difference between the positive predictions in one group versus another group. For example, if there are two age groups in the inference data for which you need to predict hospital readmissions, DPPL measures whether one age group is predicted to have more readmissions than the other.

- **Difference in Conditional Outcome (DCO)**: This metric measures the actual values for different groups of people in the inference dataset, compared to the predicted values. For example, let's assume there are two groups of people in the inference dataset: male and female. The model is trying to predict whether their health insurance claim would be accepted or rejected. This metric looks at the *actual* approved claims compared to rejected applications for both male and female applicants in the training dataset and compares it with the *predicted* values from the model in the inference dataset.

- **Recall Difference** (**RD**): Recall is a measure of a model's ability to predict a positive outcome in cases that are actually positive. RD measures the difference between the recall values of multiple groups with the inference data. If the recall is higher for one group of people in the inference dataset compared to another group, the model may be biased toward the first group.

- **Difference in Label Rates** (**DLR**): Label rate is a measure of the true positive versus predicted positive (label acceptance rate) or a measure of true negative versus predicted negative (label rejection rate). The difference between this measure across different groups in the dataset is captured by the DLR metric.

These are just four of the common metrics that SageMaker Clarify computes for you post-training. For a full list of these metrics and their details, refer to the SageMaker Clarify post-training metrics list here: `https://docs.aws.amazon.com/sagemaker/latest/dg/clarify-measure-post-training-bias.html`.

SageMaker Clarify also helps explain our model prediction behavior by generating an importance report for the features. Let us now look at model explainability with SageMaker Clarify in more detail.

Explaining model predictions with SageMaker Clarify

A model explanation is essential for building trust and understanding what is influencing a model to predict in a certain way. For example, it lets you understand why a particular patient might be at a higher risk of congestive heart failure than another. Explainability builds trust in the model by making it more transparent instead of treating it as a black box. SageMaker Clarify generates feature importance for the model using Shapley values. This helps determine the contribution of each feature in the final prediction. Shapley values are standard for the determination of feature importance and compute the effects of each feature in combination with another. SageMaker Clarify uses an optimized implementation of the Kernel SHAP algorithm. To learn more about SHAP, you can refer to the following link: `https://en.wikipedia.org/wiki/Shapley_value`.

SageMaker Clarify also allows you to generate a **partial dependence plot** (**PDP**) for your model. It shows the dependence of the predicted target on a set of input features of interest. You can configure inputs for the PDP using a JSON file. The results of the analysis are stored in the `analysis.json` file and contain the information needed to generate these plots. You can then generate a plot yourself or visualize it in the SageMaker Studio interface.

In addition to Shapley values and PDP analysis, SageMaker Clarify can also generate heatmaps for explaining predictions on image datasets. The heatmaps explain how CV models classify images and detect objects in those images. In the case of image classification, the explanations consist of images, with each image showing a heatmap of the relevant pixels involved in the prediction. In the case of object detection, SageMaker Clarify uses Shapley values to determine the contribution of each feature in the model prediction. This is represented as a heatmap that shows how important each of the features in the image scene was for the detection of the object.

Finally, SageMaker Clarify can help explain prediction behavior for **natural language processing** (**NLP**) models with classification and regression tasks. In this case, the explanations help us understand which sections of the text are the most important for the predictions made by the model. You can define the granularity of the explanations by providing the length of the text segment (such as phrases, sentences, or paragraphs). This can also extend to multimodal datasets where you have text, categorical, and numerical values in the input data.

Now that we have an idea of how SageMaker Clarify helps in bias detection and explaining model predictions, let us look at how we can monitor the performance of the models once they are deployed using real-world inference data. This is enabled using SageMaker Model Monitor.

Monitoring models with SageMaker Model Monitor

SageMaker Model Monitor allows you to continuously monitor model quality and data quality for a deployed model. It can notify you if there are deviations in data metrics or model metrics, allowing you to perform corrective actions such as retraining the model. The process of monitoring models with SageMaker Model Monitor involves four essential steps:

1. Enable the endpoint to capture data from incoming requests.
2. Create a baseline from the training dataset.
3. Create a monitoring schedule.
4. Inspect the monitoring metrics in the monitoring reports.

We will go through these steps during the exercise at the end of this chapter.

In addition, you can also enable data quality and model quality monitoring in SageMaker Studio and visualize the result on the SageMaker Studio interface. Let us now look at the details of how to set up a data quality and a model quality monitoring job on SageMaker Model Monitor.

Setting up data quality monitoring

SageMaker Model Monitor can detect drift in the dataset by comparing it against a baseline. Usually, the training data is a good baseline to compare against. SageMaker Model Monitor can then suggest a set of baseline constraints and generate statistics to explore the data. These are available inside `statistics.json` and `constraints.json` files. You can also enable SageMaker Model Monitor to directly emit data metrics to CloudWatch metrics. This allows you to generate alerts from CloudWatch if there is a violation in any of the constraints you are monitoring.

Setting up model quality monitoring

SageMaker Model Monitor can monitor the performance of a model by comparing the predicted value with the actual ground truth. It then evaluates metrics depending on the type of problem at hand. The steps to configure model quality monitoring are the same as the steps for data quality monitoring, except for an additional step to merge the ground-truth labels with the predictions. To ingest ground-truth labels, you need to periodically label the data captured by the model endpoint and upload it to a **Simple Storage Service** (**S3**) location from where SageMaker Model Monitor can read it. Moreover, there needs to be a **unique identifier** (**UID**) in the ground-truth records and a unique path on S3 where the ground truth is stored so that it can be used by SageMaker Model Monitor to access it.

It's important to note that SageMaker Model Monitor also integrates with SageMaker Clarify to enable bias-drift detection and feature attribution-drift detection on deployed models. This provides you the ability to carry out these monitoring operations on real-time data that is captured from an endpoint invocation. You can also define a monitoring schedule and a time window for when you want to monitor the deployed models.

SageMaker Clarify and Model Monitor provide you with the ability to create fair and explainable models that you can monitor in real time. Let us now use SageMaker Clarify to create a bias detection workflow and explain model predictions.

Detecting bias and explaining model predictions for healthcare coverage amounts

Bias in ML models built to predict critical healthcare metrics can erode trust in this technology and prevent large-scale adoption. In this exercise, we will start with some sample data about healthcare coverage and expenses for about 1,000 patients belonging to different demographics. We will then train a model to predict how much the coverage-to-expense ratio is for patients in different demographics. Following that, we will use SageMaker Clarify to generate bias metrics on our training data and trained model. We will also generate explanations of our prediction to understand why the model is predicting the way it is. Let's begin by acquiring the dataset for this exercise.

Acquiring the dataset

The dataset used in this exercise is synthetically generated using **Synthea**, an open source synthetic patient generator. To learn more about Synthea, you can visit the following link: `https://synthetichealth.github.io/synthea/`. Follow the next steps:

1. We use a sample dataset from Synthea for this exercise that can be downloaded from the following link: `https://synthetichealth.github.io/synthea-sample-data/downloads/synthea_sample_data_csv_apr2020.zip`.

2. The link will download a ZIP file. Unzip the file, which will create a directory called `csv` with multiple CSV files within it.

We will use only the `patients.csv` file for this exercise.

Running the Jupyter notebooks

The notebook for this exercise is saved on GitHub here:

`https://github.com/PacktPublishing/Applied-Machine-Learning-for-Healthcare-and-Life-Sciences-using-AWS/blob/main/chapter-11/bias_detection_explainability.ipynb`

The repository was cloned as part of the steps in the *Technical requirements* section. You can access the notebook from GitHub by following these steps:

1. Open the SageMaker Studio interface.

2. Navigate to the `Applied-Machine-Learning-for-Healthcare-and-Life-Sciences-using-AWS/chapter-11/` path. You should see a file named `bias_detection_explainability.ipynb`.

3. Select **New Folder** at the top of the folder navigation pane in SageMaker Studio, as shown in the following screenshot:

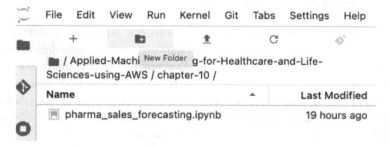

Figure 11.1 – SageMaker Studio UI showing the New Folder button

4. Name the folder data. Next, click the **Upload Files** icon on the top of the navigation pane in SageMaker Studio, as shown in the following screenshot:

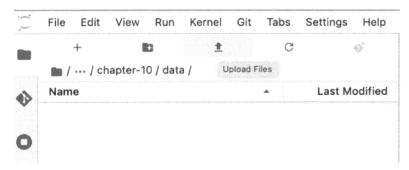

Figure 11.2 – SageMaker Studio UI showing the Upload Files button

5. Upload the patients.csv and salesdaily.csv files that you downloaded earlier.

6. Go back one folder to Chapter-11 and click on the bias_detection_explainability. ipynb file. This will open the Jupyter notebook. Follow the instructions in the notebook to complete this step.

Let us now look at the steps to visualize the reports generated in this step in SageMaker Studio. After running the notebook, you can proceed with the next section to visualize the report on SageMaker Studio.

Viewing bias and explainability reports in SageMaker Studio

After completing the steps in the preceding section, SageMaker Clarify creates two reports for you to examine. The first report allows you to look at the bias metrics for your dataset and model. The second report is the explainability report, which tells you which features are influencing the model predictions. To view the reports, follow these steps:

1. Click on **SageMaker resources** in the left navigation pane of SageMaker Studio. Make sure **Experiments and trials** is selected in the drop-down menu:

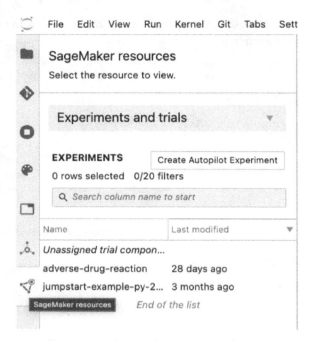

Figure 11.3 – SageMaker Studio interface showing the SageMaker resources button

2. Double-click on **Unassigned trial components** at the top of the list. In the next screen, you will see two trials. The first trial starts with **clarify-explainability** and the second starts with **clarify-bias**.

3. Double-click on the report starting with **clarify-bias**. Navigate to the **Bias report** tab, as shown in the following screenshot:

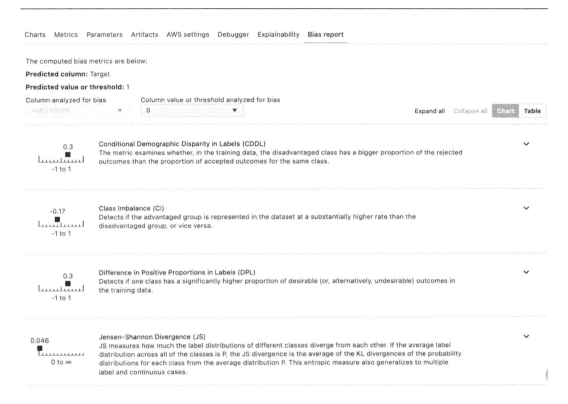

Figure 11.4 – SageMaker Clarify Bias report tab

4. Here, you can examine the bias report generated by SageMaker Clarify. It has a variety of metrics that are calculated for your dataset and model. You can expand each metric and read about what the scores might mean for your dataset and model. Spend some time reading about these metrics and assess whether the data we used has any bias in it. Remember to change the **AGEGROUPS** values using the top drop-down menu.

5. You can also view a summary of all metrics in a tabular format by clicking on the **Table** button in the top-right corner, as shown in the following screenshot:

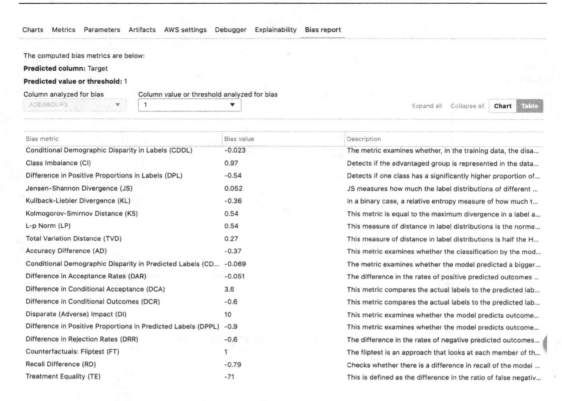

Figure 11.5 – Summary of bias metrics in SageMaker Studio

6. Next, let's look at the explainability report. Follow *steps 1* and *2* of this section to go to the screen that has the **clarify-explainability** trial. Double-click on it.

7. On the next screen, click on the **Explainability** tab at the top. This will open up the explainability report, as shown in the following screenshot:

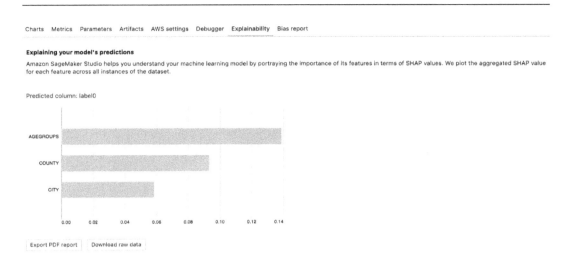

Figure 11.6 – Explainability report in SageMaker Studio

As you can see, the **AGEGROUPS** feature has the most impact on the model's prediction while the **CITY** feature has the least impact. You can also access the raw data in JSON used to generate this report and export the report into a PDF for easy sharing. This concludes the exercise.

Summary

In this chapter, we looked at some of the challenges that technologists face when implementing ML workflows in the healthcare and life sciences industry. These include regulatory challenges, security and privacy, and ensuring fairness and generalizability. We also looked at ways that organizations are addressing those challenges. We then learned about SageMaker Clarify and Model Monitor, and how they ensure fairness and provide explainability for ML models. Finally, we used SageMaker Clarify to create a bias and explainability report for a model that predicts healthcare coverage amounts.

In *Chapter 12, Understanding Current Industry Trends and Future Applications*, we will paint a picture of what to expect in the future for AI in healthcare. We will look at some newer innovations in this field and conclude by summarizing some trends.

12

Understanding Current Industry Trends and Future Applications

We have now covered different ways in which ML is making a difference for healthcare and life sciences organizations. With the help of examples in this book, you have seen that ML is more present than you may have thought and is having more impact on the way we live our lives than you may have believed. We have also explored the role AWS is playing in this transformation, particularly how the services from the AWS ML stack are making it easy for healthcare and life sciences customers to innovate at scale. From SageMaker, which lets you build, train, deploy, monitor, and operationalize ML models, to Comprehend Medical, which allows you to extract meaningful information from unstructured clinical records using pretrained models, the services cater to both experienced power users (such as research scientists) and people who are not very familiar with ML and are just getting started. The depth and breadth of these services make them applicable to all stages of the ML life cycle and help democratize the use of AI in healthcare. We also understand that there are some challenges we need to tackle. It's not easy but there are ways to address challenges with the right knowledge and capabilities.

The question now is, where do we go from here? What will the next decade hold for AI in healthcare? What new capabilities will ML provide to our healthcare organizations and what new breakthroughs will it enable? Well, no one can see into the future, but there are some trends to look out for. Advancements in research have already led to continually better-performing ML models that beat the previous **state of the art** (**SOTA**). Better instances have led to faster training and inference times that are breaking previous performance benchmarks. Thanks to **Moore's law** (`https://en.wikipedia.org/wiki/Moore%27s_law`), we now have supercomputers in the palm of our hands! We are at the cusp of something truly exciting as we face the inevitable: the merging of biology and technology.

In this chapter, we will look at some trends that might give us a clue as to where the applications of AI in healthcare and life sciences are headed. We will cover some examples of innovative use of AI in healthcare and also some technological advancements that are enabling those innovations to occur. Next, we will look at some future applications. Consider them as early experiments (at the time of writing this book) that have shown promise. We'll look at the following sections:

- Key factors influencing advancements of AI in healthcare and life sciences
- Understanding current industry trends in the application of AI for healthcare and life sciences
- Surveying the future of AI in healthcare
- Concluding thoughts

Key factors influencing advancements of AI in healthcare and life sciences

While there are multiple factors that influence advances in AI, here are some of the key factors that are having a direct impact on the healthcare and life sciences industry. Let's take a closer look at them.

Availability of multimodal data

One key dependency for any AI algorithm is the availability and access to high-quality labeled datasets. An algorithm that doesn't get exposed to enough real-world data points cannot be expected to predict true values in all scenarios. This affects the generalizability of the models and may make them biased and unfair. However, the open sharing of healthcare data is prohibited. This data contains the **protected health information** (PHI) of patients and is bound by multiple privacy and security regulations. Moreover, datasets that represent only a single modality can be constrained in the amount of information that they can capture or pass to the ML model. For instance, a physician who makes a decision about a patient's medical diagnosis gets their information from multiple modalities of data. They may be looking at the patient's imaging results, reading data from the historical test results stored in the EHR system, talking to the patient in person, and getting cues about their condition from conversations. All these data points are derived from different modalities. Similarly, for ML models to better reflect the real world, they need to be exposed to information from different data modalities.

Fortunately, we have seen a positive trend in the availability of real-world datasets in healthcare and life sciences. Multimodal datasets such as the **MIMIC 3** (`https://physionet.org/content/mimiciii/1.4/`) and the **MIMIC CxR** (`https://physionet.org/content/mimic-cxr/2.0.0/`) are great examples of public resources for healthcare. For genomics, **The Cancer Genome Atlas** (**TCGA**; `http://www.tcgaportal.org/index.html`) is paving the way for researchers to get easy access to genetic information for a variety of cancer types, leading to a better understanding of these diseases and helping to find ways to detect them early. There is government support for these kinds of initiatives. For instance, the **Centers for Medicare and Medicaid Services** (**CMS**) makes a wide range of public health data available on their portal, `Data.CMS.Gov`. The FDA has made the **OpenFDA** website (`https://open.fda.gov/data/downloads/`), containing datasets for a variety of clinical trials and drug recalls. The **Registry of Open Data** (`https://registry.opendata.aws/`) and **AWS Data Exchange** (**ADX**; `https://aws.amazon.com/data-exchange/`) are capabilities available from AWS that allow consumers to easily access, use, share, and subscribe to real-world datasets.

Active learning with human-in-the-loop pipelines

Another important factor influencing the advancement of AI in healthcare is the practice of using **active learning pipelines**. Active learning is a technique that allows algorithms to learn from human behavior and apply that learning to future events. A common use of active learning is in labeling new data. A lack of labeled datasets can pose a problem for supervised learning algorithms. Moreover, if the labels are highly specialized, it becomes a limiting factor for scaling out and utilizing crowd-sourced labelers who can generate labeled records. For example, you can get anyone to label an image with common objects such as a book, table, or tree. However, for a specialized medical image that recognizes a tumor in a brain MRI, you need highly specialized surgeons.

Active learning pipelines with humans in the loop are helping close this gap. You can start with a small subset of labeled records that can be sent to a model that starts to train and learn the labels from that small subset. The same data is sent to a group of physicians who then validate the model output and have the power to override the model-generated label if it's not accurate. As new data becomes available, the process continuously subsamples labels for training a new version of the model from the labels generated by the physicians. Over a period of time, the models learn from the ground-truth labels provided by the physicians and are able to automatically label newer records. You can automate this entire pipeline on AWS using **SageMaker Ground Truth** (`https://docs.aws.amazon.com/sagemaker/latest/dg/sms.html`).

Another key reason for using human-in-the-loop capability in ML workflows for healthcare is the concept of trust and governance. People feel more comfortable when they know that a physician is validating the output of AI models and that their healthcare decisions are not being made entirely by an AI algorithm. It makes AI more acceptable and changes the public perception of it. SageMaker **Augmented AI** (**A2I**; `https://docs.aws.amazon.com/sagemaker/latest/dg/a2i-use-augmented-ai-a2i-human-review-loops.html`) allows you to easily add human in the loop to any ML workflow so humans (in this case medical professionals) can validate and certify the output of the models.

Democratization with no-code AI tools

The AI technology landscape is undergoing a similar transformation to the one that BI and data warehousing technology went through in the 2000s. The early data warehousing projects used to rely on complex **extract, transform, load** (**ETL**) queries to pull data from multiple source systems. SQL was heavily used to process and cleanse the data with pages and pages of logic embedded within stored procedures. The process of creating the data warehousing schema involved months of data modeling efforts. The querying of data warehouses was done using analytical SQL queries that had to slice and dice facts and dimensions in a performant way. In fact, performance tuning was a key aspect of any analytical application. In today's world, everything is automated. The discovery of source system schema, pulling data from them, cleansing the data, and visualizing and slicing and dicing the data can all be done with a few clicks, or at most with a few lines of code. BI projects that used to take years to deploy in production can now be potentially completed in a few months, thanks to data transformation and BI tools.

We are seeing a similar trend in AI. Deep learning networks that used to take multiple pages of code to implement can now be done with just a few lines of code, thanks to automatic ML tools utilizing **AutoML**. These tools are becoming better every year and are making it easier for data scientists to try out multiple approaches to modeling quickly. A great example of this is **AutoGluon** (`https://auto.gluon.ai/`), which allows you to create a highly accurate ML model with just a few lines of code. We also see a lot of exciting innovations happening in no-code ML tooling spaces. These tools provide an easy-to-understand user interface that automates all stages of the ML pipeline, making it easy for anyone to try it. For instance, **SageMaker Canvas** (`https://docs.amazonaws.cn/en_us/sagemaker/latest/dg/canvas-getting-started.html`) allows you to create ML models for classification, regression, and time series forecasting problems without writing a single line of code using an easy-to-understand intuitive visual interface. We are still early in this journey of no-code ML tools, but it is all set to follow the same trajectory that the BI tooling industry followed in the early 2000s. Just like no one needs to write multiple pages of SQL to get meaningful analytics from a data warehouse, we may not need to write multiple pages of Python to create a highly accurate deep learning model. So, in a way, Python is becoming the new SQL!

Better-performing infrastructure and models

As the availability of data and AI tools increases, we also need improved infrastructure to handle the large amounts of storage and compute that are needed to process and train on that data and handle the demand of a greater number of experiments being launched as a result. Deep learning algorithms that work on large volumes of data stored in formats such as free text and images can consist of billions of parameters that need to be computed in memory. They also need better-performing GPUs and high-throughput storage. Organizations working with these large models and datasets also want to stay away from managing the large infrastructure footprint, which could be complex to scale. It is not their core competency and it takes time away from researchers, who want to concentrate on building better models to solve their research problems.

The advancements made in cloud computing technology allow us to close this gap. Organizations can get access to the best-in-class compute for a variety of workloads that may be CPU-, GPU-, or memory-bound. They can be put together in a cluster or can be containerized to create highly scalable compute environments. These compute environments are supported by multiple high-throughput storage options, making it easy for researchers to customize the environment for any ML task. Moreover, these high-end infrastructure components are available to anyone without any upfront investment. You can utilize as much or as little of the infrastructure as needed and pay for only the capacity you use, making it easy for you to optimize cost and utilization. In addition to this, AWS SageMaker providers the **data parallel** and **model parallel** libraries to train large models easily using **distributed training**. To learn more about these libraries, refer to the following site: `https://docs.aws.amazon.com/sagemaker/latest/dg/distributed-training.html`.

Better focus on responsible AI

Responsible AI refers to the steps taken by organizations to design safe, secure, and fair AI technology with good intentions that creates a positive impact on society. We all understand that AI has immense potential but can also do a lot of harm when not applied correctly. Due to this technology being new, there are hardly enough imposed regulations by agencies to govern and monitor the fair use of AI in critical fields such as defense and healthcare. Big technology firms such as Amazon, Google, Facebook, and Microsoft have all been vocal about their initiatives around **responsible AI**. They closely monitor their data collection and usage policies and have proactively maintained transparency in those practices. It has led to more confidence from the people, who are willing to put their trust in these organizations in return for the immense convenience the services provide. Even government organizations are beginning to create regulatory policies around responsible AI. A good example of this is the US **Department of Defense (DoD)**. They formed a **Defense Innovation Unit (DIU)** that launched a strategic initiative in March 2020 to implement the DoD's **Ethical Principles for Artificial Intelligence** into its commercial prototyping and acquisition programs. You can find the details of these guidelines here: `https://www.diu.mil/responsible-ai-guidelines`.

The healthcare and life sciences domain is no exception to this increase in responsible approach to AI. For instance, the FDA issued an action plan titled *Artificial Intelligence and Machine Learning in Software as a Medical Device*, which outlines five action plans that the federal administration plans to take to ensure that the safety and effectiveness of **software as a medical device (SaMD)** is maintained. Moreover, every country across the globe has its own country-specific regulations around ensuring fair and responsible use of AI technology. All this has led to more healthcare organizations than ever proactively implementing responsible AI practices in their use of AI in mission-critical healthcare workloads.

Now that we have an understanding of some of the factors influencing more healthcare and life sciences organizations to implement AI in their workloads, let us look at some industry trends to watch out for.

Understanding current industry trends in the application of AI for healthcare and life sciences

AI is now more present than ever. As we saw in the previous section, the availability of better tools, easier access to technology, and the ability for everyone to try it out are some of the key factors influencing advancements in AI/ML in healthcare and life sciences. Let us now look at some trends that show how AI applications are transforming healthcare and life sciences.

Curing incurable diseases such as cancer

The traditional method of mass-producing generic medications has existed for a while now. These medications have the same formulary and have been around for years, treating the same clinical conditions in the same way. One or two of these for a pharmaceutical company earns them billions of dollars in profits. Blockbuster drugs (as they are referred to) are produced using the same methods in large quantities and are expected to have the same effect on every patient. However, research has shown that variations in individuals and their environment play a critical role in the way they respond to medications. This is especially true for diseases such as cancer. As a result, we have now seen a trend of more personalized therapeutics development that is designed for smaller groups of individuals. These therapeutics are developed in smaller batches and are more targeted, such as focusing on a protein known to cause or spread a particular type of cancer. The design of these therapies makes heavy use of AI to engineer proteins or compounds that react with a particular target and produce the desired effect. AI allows a lot of these simulations to be carried out in silico, which also improves the chances of these drugs succeeding in clinical trials, reducing the risk for a pharmaceutical organization's upfront investments. New and innovative techniques applied to therapeutics development have led to methods such as CAR-T cell-based therapies that involve using specially formulated human cells to use in the delivery of therapies to targeted sites. Better technology in labs has led to improved equipment such as sequencers that can sequence genes faster and with much more detail than before. Better software and hardware innovation has allowed us to process extensive amounts of data generated from these processes and draw conclusions from them. Advancements in protein engineering and molecular structure prediction have provided researchers with more information about how they behave in the last few years than ever before. These trends have the potential to cure previously incurable diseases and maybe even create a world free of all diseases.

Telehealth and remote care

Healthcare is more accessible than ever. Thanks to advances in technology, you can talk to your physician from anywhere in the world, with no need to spend time waiting in long lines to talk to a nurse or book an appointment. Everything is available via your mobile phone as an app or from your computer at the click of a button. This distribution of care services via the internet and telecommunication technology is collectively known as **telehealth**.

Technology allows healthcare companies to provide remote consultations, order prescriptions or tests, and even collect samples via mail without the patient ever being in a care facility such as a clinic or hospital. This trend extends into clinical trials as well. Trial participants can now be remote and distributed across different facilities instead of traveling to a particular site where the trial is being conducted. This concept of **decentralized trials** allows for better participation rates from patients who enroll voluntarily in these trials. Having the ability to remotely access your healthcare records and having your physician or nurse available on demand helps provide better access to care and promotes health equity, especially in regions of the world where access to healthcare facilities or providers may be difficult. Telehealth digitizes your healthcare visits and associates them with your clinical history, removing the need for any manual maintenance or records, which could be error prone and inefficient. Moreover, digitizing patient visits generates a treasure trove of information that is being analyzed using ML algorithms that can help process this data and generate a lot of insightful information about your overall health and well-being. It also allows healthcare providers to better monitor you to make sure you are following all recommended actions and are regular with your medications. This trend of increased use of telehealth and using remote patient monitoring is improving patient experience and providing multiple new options for AI to create an impact on healthcare delivery.

Using the Internet of Things (IoT) and robotics

Another trend that is prevalent in healthcare AI is the increased usage of connected devices. These devices have the ability to connect to a network and also have the processing power to run ML inference, for instance, an MRI machine that can perform image segmentation to highlight a tumor on a patient's brain scan in real time directly on the machine, or a glucometer that takes regular blood sugar level readings from a patient and is able to detect anomalies that may lead to increased diabetes risk. The ML models can be deployed directly on these devices and make real-time inferences on data gathered from the device. The on-device deployment of ML models is known as **edge deployment**. Edge deployments of models directly on devices improve the latency of inferences by performing **edge inference**. Since the model is available on the device, it doesn't need to reach out to an external service to generate those inferences. Moreover, edge deployments can also handle poor connectivity for these medical devices that may not have the ability to maintain a regular connection to the network. In addition to the edge inferences, the devices usually are backed by a backend data platform that can aggregate the reading or information collected from multiple such connected devices across a period of time and perform even more aggregated analysis using ML.

The trend of edge deployments and working with real-time information has led to multiple innovations in robotics. For example, in the case of healthcare, we have seen the increasing use of robotic surgery, which involves the complex use of sensors mounted on robotic arms that can carry out a procedure with utmost accuracy and precision. Sometimes, these procedures are so advanced that only a handful of surgeons across the world can perform them. Robotic surgery can help reduce the dependency on these specialized groups of surgeons and also help with training new surgeons with similar skills.

Simulations using digital twins

One of the ways in which IoT and robotics is being applied is by utilizing them to simulate actual products or processes. For example, IoT devices mounted with sensors are being used to generate computer simulations of large machinery such as wind turbines. Using these kinds of digital simulations, engineers can understand how environmental factors affect the performance of the machinery and also alter the design digitally, until an optimal design is achieved. This sort of design alteration would be extremely hard to accomplish with actual physical systems, which could take months or years to manufacture and test. This idea of simulations could also extend to processes. For instance, computer-generated simulations can tell you how to alter your disaster management processes in the event of a failure. This simulation of a process or a physical object using computer algorithms by taking into account real-world data is known as **digital twins**.

There are multiple known applications of digital twins in the healthcare industry, and the number continues to grow every year. In the life science of therapeutics design, simulated digital twins of compounds are used to determine the composition of the therapy depending on how it interacts with the target. Simulations of infection have been used to predict how an infectious disease would spread in a population and determine how to slow its spread. Disease progression models can simulate how it would affect an individual from the time the disease is detected to the time it becomes lethal. The fact that digital twins are simulated has allowed healthcare and life sciences organizations to safely run without causing risk to patients.

All data points to an increasing trend in the adoption of AI in healthcare and life sciences. Technology and biology are merging and creating new opportunities for improving care quality across the board while finding new pathways that could lead to a world free of disease. Let's now look at the future of AI in healthcare and life sciences by summarizing some nascent areas of research for healthcare AI.

Surveying the future of AI in healthcare

Researchers are continuously pushing the boundaries of what can be achieved in healthcare and life sciences organizations with the help of AI. Some of these areas are new and experimental, while others have shown promise and are in various stages of prototyping. The following sections detail some of the new and upcoming trends in the future of AI in healthcare and life sciences.

Reinforcement learning

Reinforcement learning is a technique in ML that involves an AI algorithm learning the right sequence of decisions for a problem based on trial and error. An **agent** evaluates each trial made by the algorithm based on certain rules. The correct decisions made by the algorithm are awarded by the agent, while the incorrect ones are penalized. The overall goal of the algorithm is to maximize the reward. The point to note here is that, unlike supervised learning algorithms where the initial set of correct and incorrect outputs are fed into the algorithm during training, reinforcement learning doesn't provide a data point to the algorithm to begin with. The algorithm starts with completely random decisions and learns to narrow down to the correct one with each reward or penalty it receives from the agent during multiple runs. This makes reinforcement learning great for use cases where a sequence of decisions leads to the correct or desired outcome and when there is no labeled training data available.

Reinforcement learning applications in healthcare, while still early, can have a lot of potential applications. For example, since reinforcement learning algorithms work on sequential decision making, they could be applied to a patient's long-term care plan, which can be modified as new information about the disease or the patient's condition is available. Sequential decision making can also extend to progressive disease diagnoses, such as cancer, where a reinforcement learning algorithm can project its due course of progression based on data available as the disease spreads. These initial applications have shown a lot of promise in research and may become more mainstream in the future.

Federated learning

Federated learning is a collaborative ML technique that allows you to train a model across multiple entities, each having its own sample of training data. Unlike traditional ML approaches that expect you to centralize all training data in one location, federated learning allows for data to be decentralized and located on each client without the need to share or move the data into a central server. The process trains local models for each client with the data available locally. It then shares the model weights of the trained model with the central server. The server then computes the overall metrics for the aggregated model and sends a new version of the model to each of the clients to train on. This process continues till the desired condition is satisfied.

Federated learning has the potential to increase sharing and collaboration among researchers working on common ML problems. In healthcare, open sharing of sensitive information is not possible. Hence, researchers can use federated learning to work collaboratively with other researchers without the need to share any sensitive information. Moreover, models trained on healthcare data from only one hospital have reduced generalizability. By using federated learning, the models can be exposed to datasets across multiple hospitals, giving them exposure to new patterns and signals that they can learn from. This improves the overall generalizability of the models.

Virtual reality

Virtual reality (**VR**) is a simulated environment generated using computers. It allows users to experience a virtual world. VR can mimic the real world or it can be completely different from the real world. Users of VR can immerse themselves in the virtual world using gadgets such as a VR headset, which allows users to experience the virtual world in three dimensions and provides a 360-degree view. Users may also use sensors and controllers, which give them the ability to interact with objects in the virtual world.

One of the most popular applications of VR is in gaming. The global VR gaming market size was $11.56 billion in 2019 and is expected to grow by about 30% from 2020 to 2027.

While the gaming industry is the leader in the adoption of VR, it has valid applications in other industries as well. For example, the media and entertainment industry uses VR videos to create immersive 360-degree videos for their consumers, the education industry can create VR classrooms where students in remote learning situations can attend classes and feel more connected with their teachers and peers, and the real estate industry uses VR to create virtual tours of properties that clients could experience from the comfort of their homes. Even the healthcare industry has seen interesting applications of VR. For instance, VR assistants can answer medical questions for patients who suffer from common problems before routing patients with serious conditions to medical professionals, and medical interns can use VR to learn how to conduct complex procedures in a simulated environment that is safe but still similar to the real world. The fitness industry is increasingly adopting VR to allow users to work out from the comfort of their homes by creating real-life gymnasium experiences in the virtual world. Individuals who were hesitant to travel to the gym due to lack of time are now more willing to use VR to work out. This improves their overall health and well-being. There are countless other applications of VR in healthcare that technology companies are creating on a regular basis. The next decade is sure to see more ways in which patients experience healthcare virtually. All this leads to the increasing use of ML, which is the engine that drives these virtual environments and gets them increasingly closer to the real world.

Blockchain

Blockchain technology allows for the decentralized storage of data that cannot be owned by a single entity. The data in the blockchain is updated by a transaction that cannot be modified. This creates a decentralized public ledger of transactions that is transparent and cannot be altered. This ledger is maintained in a network of computers and is able to chain together **blocks** of information (which is where the name blockchain comes from). Each block of information consists of information about its previous blocks it is connected to. This keeps the chain traceable and also prevents it from being modified. Blockchain has multiple applications in areas where transactional information needs to be maintained for extended periods of time. A common example of the use of blockchain is in the cryptocurrency space. The immutable nature of the blockchain ledger makes it desirable for maintaining healthcare transactional information. For instance, patient history records can be maintained in the blockchain to maintain the auditability and traceability of those records. The secure nature of blockchain transactions ensures that the data is not exposed to hackers. Another common usage of blockchain in healthcare is in the supply chain space. Healthcare facilities require a lot of consumables, such as bandages, reagents, and medicines. The demand for these consumables varies over a period of time. Maintaining the supply chain information in various stages in a blockchain ensures transparency in the supply chain processes and removes inefficiencies.

Once the data is stored in the blockchain ledger, it is validated and unalterable. This makes the data ideal for training ML models. Blockchain data provides less noise and has complete information without any chances of missing data. This helps ML models become more accurate and closer to the real world. While the application of ML in blockchain is quite new, prototypes have shown a lot of promise to allow for further exploration in the future.

Quantum computing

Quantum computing is a computing technique that allows you to harness the power of **quantum mechanics** to perform computational operations. **Quantum computers**, the machines that perform these operations, use **quantum bits**, or **qubits**. Qubits can have the values 0 and 1 at the same time, which makes them quite unique when compared to a normal bit involved in classical computation, where a bit is binary (0 or 1). While there are different models of quantum computation available to quantum computers, the most common model of quantum computation is quantum circuits. Quantum circuits allow you to create a sequence of steps that are needed in your quantum computing operation, such as the initialization of qubits. While quantum computers are still behind classical computing when it comes to their performance in typical computing tasks, they have shown encouraging results for certain types of computing operations involving complex correlations between input parameters, where they have been shown to outperform classical computers.

ML can inherently take advantage of quantum computing to accelerate certain types of tasks in an ML pipeline. For example, quantum embeddings (instead of traditional embeddings) can be used to train quantum neural networks, which have advantages over traditional neural networks trained on classical embeddings. Moreover, applying quantum computing to particular problems in healthcare and life sciences that require large amounts of data processing involving correlative search parameters could have an advantage over traditional high-performance computing techniques. One potential application of quantum computing in life sciences could be in the early stages of drug discovery, where billions of compounds need to be evaluated for desirable properties. In the future, when quantum computing is more accessible and available to researchers and developers, we will certainly see more applications of its unique capability in applied AI/ML for healthcare and life sciences problems.

While no one has seen the future, these are certainly new areas of innovation that will define the AI landscape for healthcare and life sciences. I will surely be keeping a close eye on these to see how it evolves.

Concluding thoughts

The US spent a record 20% of its GDP on healthcare in 2020. While the costs are high, it is also promising to note that a large portion of that budget has been used for the modernization of the healthcare and life sciences technology landscape. Overhauling any existing domain with technology does require some upfront investment and time. The long-term return on these investments is expected to overshadow the costs in the shorter term. For example, investments like these have led to a decrease in cancer deaths in the US. A report published by the American Association for Cancer Research (`https://cancerprogressreport.aacr.org/wp-content/uploads/sites/2/2022/09/AACR_CPR_2022.pdf`) found that deaths from cancer have decreased by 2.3% every year between 2016 and 2019 and is on a downward trend. These statistics make me more optimistic that technology, specifically AI and ML, can have a meaningful impact on our health and wellness. Easier access to technology and the proof that these technologies have a direct impact on improving our health and well-being are going to lead to transformational change in the way healthcare is conceived. From the discovery of new therapies and drugs to its delivery to patients and tracking the effects it has on their lives, the entire value chain of healthcare and life sciences is connected and available to us to analyze using a single pane of glass. Biotech organizations that utilize advanced analytics and ML can draw out insights from this data that no one thought was possible. No wonder that according to a publication in *Nature*, the US biotech sector revenue is estimated to have grown on average >10% each year over the past decade, which is much faster than the rest of the economy (`https://www.nature.com/articles/nbt.3491`). This makes biotech one of the fastest-growing segments of the US economy, contributing $300-400 billion annually. While a lot of this may just sound like a dream, it is not. It is real. We are already seeing returns on these investments and the future is going to provide many more examples.

Summary

In this chapter, we got an understanding of some of the factors that are directly responsible for the increasing use of AI in the healthcare and life sciences industry. We also looked at some common trends that AI is influencing, from curing incurable diseases to the use of IoT, robotics, and digital twins. We also summarized a few topics to watch out for. These topics are new, with limited real-world data around their successes, but they have a lot of potential for impact.

This chapter concludes our journey. The guidance in this book summarizes the years of learning that I have personally gone through and continue to experience. Through the chapters in this book, I have tried to summarize the use of AI/ML in healthcare and life sciences in a structured and accessible manner. The practical, applied ML implementation examples using AWS services should help articulate the role AWS is playing in making AI more accessible to healthcare and life sciences organizations. I am fortunate to have a front-row seat on this innovation journey and I am in awe. I hope I have passed this feeling on to you. With the knowledge you now have, I encourage you to continue on this learning path and apply your creativity and technical abilities to create new applications for healthcare and life sciences that utilize AI/ML. This is just the beginning.

Index

Packt.com

Subscribe to our online digital library for full access to over 7,000 books and videos, as well as industry leading tools to help you plan your personal development and advance your career. For more information, please visit our website.

Why subscribe?

- Spend less time learning and more time coding with practical eBooks and Videos from over 4,000 industry professionals
- Improve your learning with Skill Plans built especially for you
- Get a free eBook or video every month
- Fully searchable for easy access to vital information
- Copy and paste, print, and bookmark content

Did you know that Packt offers eBook versions of every book published, with PDF and ePub files available? You can upgrade to the eBook version at packt.com and as a print book customer, you are entitled to a discount on the eBook copy. Get in touch with us at customercare@packtpub.com for more details.

At www.packt.com, you can also read a collection of free technical articles, sign up for a range of free newsletters, and receive exclusive discounts and offers on Packt books and eBooks.

Other Books You May Enjoy

If you enjoyed this book, you may be interested in these other books by Packt:

Machine Learning in Biotechnology and Life Sciences

Saleh Alkhalifa

ISBN: 9781801811910

- Get started with Python programming and Structured Query Language (SQL)
- Develop a machine learning predictive model from scratch using Python
- Fine-tune deep learning models to optimize their performance for various tasks
- Find out how to deploy, evaluate, and monitor a model in the cloud
- Understand how to apply advanced techniques to real-world data
- Discover how to use key deep learning methods such as LSTMs and transformers

Applied Machine Learning Explainability Techniques

Aditya Bhattacharya

ISBN: 9781803246154

- Explore various explanation methods and their evaluation criteria
- Learn model explanation methods for structured and unstructured data
- Apply data-centric XAI for practical problem-solving
- Hands-on exposure to LIME, SHAP, TCAV, DALEX, ALIBI, DiCE, and others
- Discover industrial best practices for explainable ML systems
- Use user-centric XAI to bring AI closer to non-technical end users
- Address open challenges in XAI using the recommended guidelines

Packt is searching for authors like you

If you're interested in becoming an author for Packt, please visit authors.packtpub.com and apply today. We have worked with thousands of developers and tech professionals, just like you, to help them share their insight with the global tech community. You can make a general application, apply for a specific hot topic that we are recruiting an author for, or submit your own idea.

Share Your Thoughts

Now you've finished *Applied Machine Learning for Healthcare and Life Sciences using AWS*, we'd love to hear your thoughts! Scan the QR code below to go straight to the Amazon review page for this book and share your feedback or leave a review on the site that you purchased it from.

https://packt.link/r/1-804-61021-6

Your review is important to us and the tech community and will help us make sure we're delivering excellent quality content.

Download a free PDF copy of this book

Thanks for purchasing this book!

Do you like to read on the go but are unable to carry your print books everywhere? Is your eBook purchase not compatible with the device of your choice?

Don't worry, now with every Packt book you get a DRM-free PDF version of that book at no cost.

Read anywhere, any place, on any device. Search, copy, and paste code from your favorite technical books directly into your application.

The perks don't stop there, you can get exclusive access to discounts, newsletters, and great free content in your inbox daily

Follow these simple steps to get the benefits:

1. Scan the QR code or visit the link below

https://packt.link/free-ebook/9781804610213

2. Submit your proof of purchase
3. That's it! We'll send your free PDF and other benefits to your email directly